T0178417

Communications
in Computer and Information Science 1945

Editorial Board Members

Rationale

The CCIS series is devoted to the publication of proceedings of computer science conferences. Its aim is to efficiently disseminate original research results in informatics in printed and electronic form. While the focus is on publication of peer-reviewed full papers presenting mature work, inclusion of reviewed short papers reporting on work in progress is welcome, too. Besides globally relevant meetings with internationally representative program committees guaranteeing a strict peer-reviewing and paper selection process, conferences run by societies or of high regional or national relevance are also considered for publication.

Topics

The topical scope of CCIS spans the entire spectrum of informatics ranging from foundational topics in the theory of computing to information and communications science and technology and a broad variety of interdisciplinary application fields.

Information for Volume Editors and Authors

Publication in CCIS is free of charge. No royalties are paid, however, we offer registered conference participants temporary free access to the online version of the conference proceedings on SpringerLink (http://link.springer.com) by means of an http referrer from the conference website and/or a number of complimentary printed copies, as specified in the official acceptance email of the event.

CCIS proceedings can be published in time for distribution at conferences or as postproceedings, and delivered in the form of printed books and/or electronically as USBs and/or e-content licenses for accessing proceedings at SpringerLink. Furthermore, CCIS proceedings are included in the CCIS electronic book series hosted in the SpringerLink digital library at http://link.springer.com/bookseries/7899. Conferences publishing in CCIS are allowed to use Online Conference Service (OCS) for managing the whole proceedings lifecycle (from submission and reviewing to preparing for publication) free of charge.

Publication process

The language of publication is exclusively English. Authors publishing in CCIS have to sign the Springer CCIS copyright transfer form, however, they are free to use their material published in CCIS for substantially changed, more elaborate subsequent publications elsewhere. For the preparation of the camera-ready papers/files, authors have to strictly adhere to the Springer CCIS Authors' Instructions and are strongly encouraged to use the CCIS LaTeX style files or templates.

Abstracting/Indexing

CCIS is abstracted/indexed in DBLP, Google Scholar, EI-Compendex, Mathematical Reviews, SCImago, Scopus. CCIS volumes are also submitted for the inclusion in ISI Proceedings.

How to start

To start the evaluation of your proposal for inclusion in the CCIS series, please send an e-mail to ccis@springer.com.

Feng Wu · Xuanjing Huang · Xiangnan He ·
Jiliang Tang · Shu Zhao · Daifeng Li · Jing Zhang
Editors

Social Media Processing

11th Chinese National Conference, SMP 2023
Anhui, China, November 23–26, 2023
Proceedings

 Springer

Editors
Feng Wu
University of Science and Technology
of China
Anhui, China

Xiangnan He
University of Science and Technology
of China
Anhui, China

Shu Zhao
Anhui University
Anhui, China

Jing Zhang
Renmin University of China
Beijing, China

Xuanjing Huang
Fudan University
Shanghai, China

Jiliang Tang
Michigan State University
East Lansing, MI, USA

Daifeng Li
Sun Yat-sen University
Guangzhou, China

ISSN 1865-0929   ISSN 1865-0937 (electronic)
Communications in Computer and Information Science
ISBN 978-981-99-7595-2   ISBN 978-981-99-7596-9 (eBook)
https://doi.org/10.1007/978-981-99-7596-9

This Springer imprint is published by the registered company Springer Nature Singapore Pte Ltd.
The registered company address is: 152 Beach Road, #21-01/04 Gateway East, Singapore 189721, Singapore

Paper in this product is recyclable.

Preface

This volume contains the papers presented at SMP 2023: the National Social Media Processing Conference held on November 23–26, 2023, in Anhui.

The Social Media Processing Special Committee of the Chinese Information Society of China organizes SMP. SMP focuses on scientific research centered on social media processing. It provides a broad platform for disseminating the latest academic research and technological achievements in social media processing, aiming to build an industry-academia-research ecosystem in the field of social media processing. It strives to become a trendsetter in social media processing in China and even globally.

SMP was founded in 2012 and is held annually (paused for a year in 2021 due to external reasons). It has now become an important academic event in the field of social media processing. The theme of this year is Social Media Meets Big Model. Enclosing this theme, the conference scheduled various activities, including keynotes, academic workshops, poster presentations, etc.

As for peer-reviewed papers, 88 submissions were received, mainly covering the following areas,

- Large Models and Social Media
- Social Media Mining and Prediction
- Network Representation Learning, Graph Neural Network Theory and Applications
- Social Recommendation
- Natural Language Processing for Social Media
- Social Multimedia Computing

During the reviewing process, each submission was assigned to at least three Program Committee members. The committee accepted 39 full papers (16 papers in English). The CCIS volume contains revised versions of 16 English full papers.

The hard work and close collaboration of a number of people contributed to the success of this conference. We want to thank the Organizing Committee and Program Committee members for their support, and the authors and participants who are the primary reason for the success of this conference. We also thank Springer for their trust and for publishing the proceedings of SMP 2023.

Finally, we appreciate the sponsorships from Huashang Securities and iFlytek as chief sponsors, Zhipu and Hefei Municipal People's Government as diamond sponsors, Huawei, Baidu, Sina, Midu, Ping An, NiuTrans, Xiaomi, and ModelBest as platinum sponsors, Bayou Technology, Qingbo Big Data, KeAi, Yunfu Technology, Paratera and Alex Masen as gold sponsors.

November 2023

Feng Wu
Xuanjing Huang
Xiangnan He
Jiliang Tang
Shu Zhao
Daifeng Li
Jing Zhang

Organization

General Chairs

Feng Wu   University of Science and Technology of China,
                             China
Xuanjing Huang   Fudan University, China

Program Committee Chairs

Xiangnan He   University of Science and Technology of China,
                             China
Jiliang Tang   Michigan State University, USA
Shu Zhao   Anhui University, China

Local Organizing Committee Chairs

Xiao Sun   Hefei University of Technology, China
Fulan Qian   Anhui University, China

Publicity Chairs

Xiang Wang   University of Science and Technology of China,
                             China
Xu Chen   Renmin University of China, China

Sponsorship Chairs

Binyang Li   Institute of International Relations, China
Fuli Feng   University of Science and Technology of China,
                                    China

Evaluation Chairs

Huaiyu Wan   Beijing Jiaotong University, China
Liang Yang   Dalian University of Technology, China

Tutorial Chairs

Le Wu   Hefei University of Technology, China
Lun Zhang   Beijing Normal University, China

Forum Chairs

Tong Xu   University of Science and Technology of China,
                                    China
Zhongyu Wei   Fudan University, China

Publication Chairs

Daifeng Li   Sun Yat-sen University, China
Jing Zhang   Renmin University of China, China

Doctoral Academic Forum Chairs

Junwen Duan   Central South University, China
Jiarong Xu   Fudan University, China

Industry Track Chairs

Cunchao Tu	PowerLaw, China
Xuqi Zhu	Qingbo Intelligent, China

Poster Chairs

Yang Li	Northeast Forestry University, China
Jie Chen	Anhui University, China

Program Committee

Aiwen, Jiang	Jiangxi Normal University, China
Bin, Li	Wuhan University, China
Bingbing, Xu	Institute of Computing Technology, Chinese Academy of Sciences, China
Binyang, Li	University of International Relations, China
Bo, Xu	Dalian University of Technology, China
Changxuan, Wan	Jiangxi University of Finance and Economics, China
Chen, Jie	Anhui University, China
Chen, Qian	Shanxi University, China
Cheng, Zheng	Anhui University, China
Chengzhi, Zhang	Nanjing University of Science and Technology, China
Chenliang, Li	Wuhan University, China
Daifeng, Li	Baidu, China
Dexi, Liu	Jiangxi University of Finance and Economics, China
Di, Zhao	Dalian Minzu University, China
Dong, Liu	Henan Normal University, China
Dong, Zhou	Guangdong University of Foreign Studies, China
Dongyu, Zhang	Dalian University of Technology, China
Fangzhao, Wu	Microsoft Research Asia, China
Fei, Xiong	Beijing Jiaotong University, China
Fulan, Qian	Anhui University, China
Fuli, Feng	University of Science and Technology of China, China
Guo, Chen	Nanjing University of Science and Technology, China

Guojie, Song	Peking University, China
Hong, Huang	TUGoe, China
Hongfei, Lin	Dalian University of Technology, China
Huaiyu, Wan	Beijing Jiaotong University, China
Huawei, Shen	Chinese Academy of Sciences, China
Huiting, Liu	Anhui University, China
Jia, Liu	Xihua University, China
Jiali, Zuo	Jiangxi Normal University, China
Jian, Jiang	Wuhan Textile University, China
Jian, Liao	Shanxi University, China
Jian, Zhan	Lanzhou University, China
Jiancan, Wu	University of Science and Technology of China, China
Jiarong, Xu	Fudan University, China
Jiawei, Chen	Zhejiang University, China
Jifan, Yu	Tsinghua University, China
Jiliang, Tang	Michigan State University, USA
Jing, Zhang	Renmin University of China, China
Jiuxin, Cao	Southeast University, China
Jufeng, Yang	Nankai University, China
Junchan, Zhao	Hunan University of Technology and Business, China
Junru, Chen	Zhejiang University, China
Junwen, Duan	Central South University, China
Kaiyu, Li	York University, Canada
Kan, Xu	Dalian University of Technology, China
Le, Wu	Hefei University of Technology, China
Lei, Guo	Shandong Normal University, China
Li, Wang	Taiyuan University of Technology, China
Liang, Yang	Dalian University of Technology, China
Lizhou, Feng	Jilin University, China
Lu, Ren	Shandong Technology and Business University, China
Lun, Zhang	Beijing Normal University, China
Meng, Jia-Na	Dalian University of Technology, China
Min, Zhang	Tsinghua University, China
Min, Zhou	Beijing Normal University, China
Mingwen, Wang	Jiangxi Normal University, China
Na, Ta	Renmin University of China, China
Piji, Li	Nanjing University of Aeronautics and Astronautics, China
Qi, Huang	Jiangxi Normal University, China

Qingchao, Kong	Institute of Automation, Chinese Academy of Sciences, China
Qizhi, Wan	Jiangxi University of Finance and Economics, China
Rao, Gaoqi	Beijing Language and Culture University, China
Ronggui, Huang	Fudan University, China
Ruifeng, Xu	Harbin Institute of Technology, China
Ruihua, Qi	Dalian University of Foreign Languages, China
Saike, He	Beijing University of Posts and Telecommunications, China
Shanshan, Wang	University of Jinan, China
Shi, Feng	Northeastern University, China
Shichang, Sun	Dalian Minzu University, China
Shu, Zhao	Anhui University, China
Shuxin, Yang	Jiangxi University of Science and Technology, China
Sihao, Ding	University of Science and Technology of China, China
Tao, Jia	Southwest University, China
Tao, Lian	Taiyuan University of Technology, China
Ting, Wang	National University of Defense Technology, China
Tong, Mo	Peking University, China
Tong, Xu	University of Science and Technology of China, China
Wei, Wei	Huazhong University of Science and Technology, China
Wei, Zhang	East China Normal University, China
Weiyun, Qiu	Shandong University, China
Wenji, Mao	Institute of Automation, Chinese Academy of Sciences, China
Wenjun, Jiang	Hunan University, China
Wu, Xiaolan	Nanjing University of Science and Technology, China
Xia, Li	Guangdong University of Foreign Studies, China
Xiang, Ao	Institute of Computing Technology, Chinese Academy of Sciences, China
Xiang, Wang	National University of Singapore, Singapore
Xiangju, Li	Shandong University of Science and Technology, China
Xiangnan, He	University of Science and Technology of China, China
Xiao, Ding	Harbin Institute of Technology, China

Xiao, Sun	Hefei University of Technology, China
Xiao, Sun	Institute of Artificial Intelligence, Hefei Comprehensive National Science Center, China
Xiaohui, Han	Qilu University of Technology, China
Xiaoke, Xu	Dalian Minzu University, China
Xiaoliang, Chen	Xihua University, China
Xin, Huang	Hong Kong Baptist University, China
Xin, Wang	Jilin University, China
Xin, Xin	Shandong University, China
Xu, Chen	Renmin University of China, China
Xue-Qiang, Zeng	Jiangxi Normal University, China
Yang, Yang	Zhejiang University, China
Yanli, Li	Xihua University, China
Yanling, Wang	Renmin University of China, China
Yifei, Sun	Zhejiang University, China
Yijia, Zhang	Dalian Maritime University, China
Ying, Wang	Jilin University, China
Yongbin, Liu	Tsinghua University, China
Yongjun, Li	Northwestern Polytechnical University, China
Yongqing, Wang	Institute of Automation, Chinese Academy of Sciences, China
Yu, Houqiang	Wuhan University, China
Yuan, Gao	University of Science and Technology of China, China
Yuanyuan, Sun	Dalian University of Technology, China
Yufeng, Diao	Inner Mongolian Minzu University, China
Yuhua, Li	Huazhong University of Science and Technology, China
Yujun, Zhou	
Yuyue, Zhao	University of Science and Technology of China, China
Zhaochun, Ren	Leiden University, The Netherlands
Zhenyu, Wang	South China University of Technology, China
Zhongying, Zhao	Shandong University of Science and Technology, China
Zhongyu, Wei	Fudan University, China
Zhumin, Chen	Shandong University, China
Zhuoren, Jiang	Zhejiang University, China

Contents

DABP: A Domain Augmentation and Bidirectional Stack-Propagation Model for Task-Oriented NLU

Shizhan Lan[1,2], Yuehao Xiao[1], and Zhenyu Wang[1(✉)]

[1] South China University of Technology, Guangzhou 510006, China
`sexiaoyh@mail.scut.edu.cn, wangzy@scut.edu.cn`
[2] China Mobile Guangxi Branch Co., Ltd., Nanning 530012, China
`lanshizhan@gx.chinamobile.com`

Abstract. Natural language understanding (NLU) is the key part of task-oriented dialogue systems. Nowadays, most existing task-oriented NLU models use pre-trained models (PTMs) for semantic encoding, but those PTMs often perform poorly on specific task-oriented dialogue data due to small data volume and lack of domain-specific knowledge. Besides that, most joint modeling models of slot filling and intention detection only use a joint loss function, or only provides a one-way semantic connection, which fails to achieve the interaction of information between the two tasks at a deep level. In this paper, we propose a Domain Augmentation and Bidirectional Stack Propagation (DABP) model for NLU. In the proposed model, we use the masked language model (MLM) task and the proposed part-of-speech tagging task to enhance PTMs with domain-specific knowledge include both implicit and explicit. Besides that, we propose a bidirectional stack-propagation mechanism to propagate the information between the two tasks. Experimental results show that the proposed model can achieve better performance than the state-of-the-art models on the ATIS and SNIPS datasets.

Keywords: Natural Language Understanding · Domain Augmentation · Bidirectional Stack-propagation

1 Introduction

Task-oriented dialogue system aims to complete certain needs in a certain field for users through language communication with users. Compared with open-domain dialogue systems whose main goal is to maximize user engagement [1], task-oriented dialogue systems focus more on completing certain tasks in one or more domains, and are built on a structured ontology that defines domain knowledge. The NLU module is the basis of the task-oriented dialogue system based on the pipeline method. NLU can be viewed as consisting of two subtasks: slot filling and intent detection. Slot filling is a sequence labeling task that involves annotating key information using BIO encoding [2], while intent detection is a classification task.

F. Wu et al. (Eds.): SMP 2023, CCIS 1945, pp. 1–13, 2024.
https://doi.org/10.1007/978-981-99-7596-9_1

Currently, the majority of mainstream NLU methods utilize pre-trained models for semantic encoding and use the hidden layer output as input for slot filling and intent detection tasks [3]. However, these pre-trained models are trained on large-scale non-target domain data, resulting in over-fitting and unsatisfactory performance in the target domain due to the small data size [4]. In this paper, we propose a solution to this issue by using a pre-training model to continue pre-training on the downstream task dataset. Our approach involves incorporating explicit knowledge-enhanced part-of-speech tagging (POS Tagging) task while using the MLM task. We jointly optimize the losses of these two pre-training tasks to achieve the best results.

Moreover, previous NLU models often treated slot filling and intent detection separately, without taking into account their close relationship. While some joint modeling models have been proposed recently [5], most of them only use a joint loss function that fails to establish deep-level interactions between the two tasks [6], or provide a one-way semantic connection [7]. In this paper, we propose a bidirectional stack propagation-based NLU model that establishes a bidirectional semantic interaction relationship between slot filling and intent detection. We use the hidden layer containing intent information as the input for the slot filling task and the context vector containing slot value information as the input for the intent detection task.

We evaluate our model on the ATIS and Snips datasets, and experimental results demonstrate its competitive performance in slot filling and intent detection tasks.

2   Problem Definition

The NLU module consists of two tasks: slot filling and intent detection, whose goal is to train a probability distribution model to map a user utterance $\mathbf{x} = \{x_1, x_2, \ldots, x_n\}$ to a slot label sequence $\mathbf{s} = \{s_1, s_2, \ldots, s_n\}$ and an intent category y. This is achieved by using a dataset $D = \{(\mathbf{x}_1, \mathbf{s}_1, y_1), \ldots, (\mathbf{w}_n, \mathbf{s}_n, y_n)\}$ for training.

3   The DABP

The whole framework of proposed DABP is shown in Fig. 1. We use a pre-trained model enhanced by masking language modeling tasks and part-of-speech tagging tasks on domain-specific datasets to obtain semantic feature representations of input sequences. Then the two-way stack propagation network performs joint modeling of slot filling and intent detection on the semantic feature representation. By directly connecting the two tasks at the semantic level, two complementary feature vectors are used to predict the intent and slot value respectively, and the intent Feature information and slot value feature information complement and promote each other.

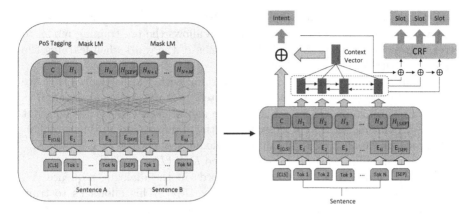

Fig. 1. The framework of DABP

3.1 Domain Augmented Pre-trained Model

The pre-trained model we use is the uncased version of BERT [8]. In order to bring out the potential of the pre-trained model in task-oriented dialogue, two domain augmentation methods are used, namely the MLM task based on whole word masking (WWM) and the part-of-speech tagging task. These two tasks are based on the input sequence of the pre-training model, which can effectively use the contextual implicit information of the dialogue sentence and the explicit information of the language tagging, so that the pre-training model can be augmented in understanding the content of the dialogue sentence in a specific field.

MLM Task Based on WWM. MLM is an unsupervised pre-training method derived from BERT. Its purpose is to let the pre-training model learn the contextual implicit semantic information in the input sequence. In order to train a bidirectional language feature representation, the task directly randomly covers a certain proportion of words in the input, and then predicts these covered parts. This chapter uses the same random ratio as BERT, which is 15%. In the actual replacement operation of masked words, 80% of the cases are replaced by [MASK], 10% of the cases are replaced by random words, and the remaining 10% of the cases are kept unchanged.

In the MLM task of original BERT, the WordPiece tokenizer splits words into multiple subword tokens, with the number of tokens varying based on the length of the word. For instance, the word "bankers" may be tokenized into three subword tokens: "bank", "##er", and "##s", while "banks" may be tokenized into two subword tokens: "bank" and "##s". However, such tokenization only considers the local semantics of the subword tokens, while ignoring the complete semantics of the original vocabulary.

To address this issue, we use the WWM method, where if any part of a word is masked, the entire word is masked instead of just the masked part. For example, in the above illustration, the whole word "bankers" will be masked if any of its subword tokens are masked. This approach allows the pre-training model to better utilize the complete semantic information of the word when learning the implicit semantic information of the context.

To begin with, our approach involves working with text sequences that are composed of task-specific dialogues, represented as $A = \{A_1, \ldots, A_a\}$ and $B = \{B_1, \ldots, B_b\}$ and so on. These sequences are concatenated to create the input sequence X, which is then preprocessed using the Stanford CoreNLP natural language processing toolkit [9] to obtain the dialogue sequence $X' = \{X'_1, \ldots, X'_n\}$ and the corresponding part-of-speech sequence $D = \{D_1, \ldots, D_n\}$.

Next, we randomly mask 15% of the words in the dialogue sequence X', with the number of masked words $k = \lfloor N \times 15\% \rfloor$. This allows us to train the model to better handle missing information and improve its robustness to noise. The masked text sequence X' is then passed through the embedding layer of the pre-trained model to obtain the input vector $H^{(0)}$. This input vector is composed of character embedding vectors, segment embedding vectors, and position embedding vectors.

$$X = [\text{CLS}]A_1 \ldots A_n[\text{SEP}]B_1 \ldots B_m[\text{SEP}] \ldots \tag{1}$$

$$X' = \text{WordSegmentation}(X) \tag{2}$$

$$D = \text{PoSTagging}(X) \tag{3}$$

$$H^{(0)} = \text{Embedding}(X') \tag{4}$$

Then, contextual semantic representation $H^{(L)} \in \mathbb{R}^{N \times d}$ is derived by passing the input sequence through a stack of L Transformer networks. The resulting representation captures the contextual information of the input, where N denotes the maximum sequence length and d denotes the dimensionality of the Transformer network's hidden layer.

$$H^{(i)} = \text{Transformer}\left(H^{(i-1)}\right), i \in \{1, \ldots, L\} \tag{5}$$

Since it is only necessary to predict the words replaced by the masked language modeling task, after obtaining the contextual semantic representation H^L, a subset of the masked positions is selected to form a representation of the masked words, denoted as $H^m \in \mathbb{R}^{k \times d}$, where k is the number of masked words.

The representation H^m is then projected into the vocabulary space to predict the probability distribution over the entire vocabulary $\mathbb{V}$. To achieve this, the word embedding matrix $W^e \in \mathbb{R}^{|V| \times d}$, which has the same dimensionality as the hidden layer, is used for projection.

$$p_i = H_i^m W^{e^T} + b \tag{6}$$

Finally, a standard cross-entropy loss function is used to optimize the masked language modeling task.

$$\mathcal{L} = -\frac{1}{M} \sum_{i=1}^{M} y_i \log p_i \tag{7}$$

PoS Tagging. The primary objective of language modeling for large-scale dialogue datasets is to capture both explicit and implicit knowledge automatically. Explicit knowledge can be easily expressed and shared through words, numbers, and other forms of structured data. In contrast, implicit knowledge is subjective, difficult to structure, and often manifests as a deep understanding of language.

Traditional pre-training tasks have mainly focused on mining implicit knowledge from large-scale datasets, while neglecting the importance of explicit knowledge. To address this gap, we propose a part-of-speech tagging task that enables the model to capture explicit knowledge, enhance language representation, and improve language understanding and reasoning capabilities. By incorporating this task into the augmentation process, the model can learn to identify the grammatical structure of sentences and extract valuable information about word usage, which can aid in downstream tasks that require explicit knowledge.

To perform the part-of-speech tagging task, we use the part-of-speech sequence $D = \{D_1, \ldots, D_n\}$ obtained through the above processing as the gold standard. During training, after masking a random portion of the input sequence, we project the corresponding contextual semantic representation H of the [CLS] tag in the preprocessed input sequence X' to predict the part of speech of the masked portion.

$$p = \mathrm{softmax}\left(HW^{\mathrm{s}} + b^{\mathrm{s}}\right) \tag{8}$$

Among them, $W^{\mathrm{s}} \in \mathbb{R}^{d \times |S|}$ is the weight matrix of the part-of-speech tag prediction layer, $b^{\mathrm{s}} \in \mathbb{R}^{1 \times |S|}$ is the bias vector, $|S|$ is the number of categories of part-of-speech tags.

To optimize the part-of-speech tagging task, we use a standard cross-entropy loss function, which measures the discrepancy between the predicted probability distribution over the part-of-speech tag categories and the true distribution.

During joint training, we combine the losses of the masked language modeling task and the part-of-speech tagging task. Specifically, we compute the sum of the losses of the two tasks and use it as the loss function for the final pre-training model domain data training enhancement task. This joint training approach enables the model to capture both explicit and implicit knowledge from the input sequence, leading to improved performance on downstream tasks that require a deep understanding of language.

$$\mathcal{L} = \mathcal{L}_{\mathrm{mlm}} + \mathcal{L}_{\mathrm{pos\text{-}tagging}} \tag{9}$$

After training to convergence, the BERT augmented by domain data is used as the pre-training model used in the bidirectional stack propagation network.

3.2 Bidirectional Stack Propagation Network

In traditional nlu models, slot filling and intent detection are often treated as two independent tasks. However, in task-oriented dialogues, there is a close relationship between these two tasks. For example, for a given intent, the corresponding slot values often have certain range constraints, while for a given slot value, it can provide inspirations for intent detection. Therefore, there exists a certain bidirectional dependency between slot filling and intent detection.

Some researchers have begun to adopt joint models to simultaneously model these two tasks. However, most joint modeling approaches only encode user utterances into slot feature vectors and intent feature vectors, and then perform slot filling and intent detection through independent sub-networks, finally using a joint loss function to implicitly associate the two tasks. In these joint modeling approaches, the interaction between slot feature information and intent feature information only occurs on the surface level and is not deeply bidirectional.

Recently, Qin et al. [7] proposed a stack propagation model that directly uses intent information as the input for slot filling tasks, thereby leveraging the semantic correlation between the two. However, this method only provides one-way stack propagation and does not establish a bidirectional stack propagation relationship.

Based on this idea, in this section, we propose a bidirectional stack propagation network, which is based on the single-directional stack propagation idea but adds bidirectional stack propagation relationships. This approach enhances the interaction between slot feature information and intent feature information, leading to improved performance on slot filling and intent detection tasks in task-oriented dialogues.

Intent Detection Subnet. After domain augmentation, the user utterance sequence $X = \{X_1, \ldots, X_n\}$ is encoded by the pre-trained model, resulting in a representaion vector sequence $H = \{H_1, \ldots, H_n\}$, where $H_i \in \mathbb{R}^d$ and d is the dimensionality of the representaion vector. Through the self-attention mechanism of the Transformer structure in the pre-trained model, the representaion vector sequence H contains semantic information of the context. Then, the representaion vector sequence H is input to a BiLSTM to obtain the hidden layer vector $H_i' \in \mathbb{R}^{d_h}$ at each time step, where d_h is the dimensionality of the hidden layer vector.

$$X = [\text{CLS}]A_1 \ldots A_j[\text{SEP}]B_1 \ldots B_k[\text{SEP}] \tag{10}$$

$$H = \text{PLM}(X) = \{H_1, \cdots, H_n\} \tag{11}$$

$$H_i' = \text{BiLSTM}\left(H_{i-1}', H_{i+1}', H_i\right) \tag{12}$$

The hidden layer vector sequence $H' = \{H_1', \ldots, H_n'\}$ output by the BiLSTM contains slot value feature information at different positions and time steps. To represent the semantic information of the entire sentence, the hidden layer vector sequence H' is input into a self-attention module to obtain a context vector C that contains the information of the entire dialogue.

In the self-attention module, the projection parameters W_Q, W_K, and W_V are used to map the hidden layer vector H' to three different feature spaces: Q, K, and V. Then, attention weights are calculated through dot-product scaling and Softmax, where the square root of the dimension of K is used for scaling to obtain a more stable normal distribution and reduce the impact of the length of K. Finally, the attention weight sequence $\alpha = \{\alpha_1, \ldots, \alpha_n\}$ and the vector V are weighted and summed to obtain the context vector $C \in \mathbb{R}^{d_h}$, where d_h is the dimensionality of the context vector.

$$C = \sum_{i=1}^{n} \alpha_i V_i \tag{13}$$

$$\alpha_i = \frac{\exp\left(\frac{Q_i^T \cdot K_i}{\sqrt{D_{K_i}}}\right)}{\sum_{j=1}^{n} \exp\left(\frac{Q_i^T \cdot K_i}{\sqrt{D_{K_i}}}\right)} = \text{Softmax}\left(\frac{Q_i^T \cdot K_i}{\sqrt{D_{K_i}}}\right) \tag{14}$$

$$Q_i = W_Q H_i' \tag{15}$$

$$K_i = W_K H_i' \tag{16}$$

$$V_i = W_V H_i' \tag{17}$$

The context vector C and the intent feature vector H_0 output by the pre-trained model are concatenated to obtain an intent-enhanced vector $I \in \mathbb{R}^{d_C + d_{H_0}}$, where d_C is the dimensionality of the context vector and d_{H_0} is the dimensionality of the feature vector. The intent-enhanced vector I is then input into the softmax activation function to obtain the intent detection result y.

$$I = [C, H_0] \tag{18}$$

$$y = \text{Softmax}(I) \tag{19}$$

Slot Filling Subnet. The intent-enhanced vector I and the hidden layer vector H_i at each time step output by the BiLSTM are concatenated to obtain the slot-enhanced vector $Z_i \in \mathbb{R}^{d_I + d_{H_i}}$. The slot-enhanced vector Z is then normalized through Softmax to obtain a feature matrix $P \in \mathbb{R}^{n \times k}$, where k represents the number of slot value categories.

To better utilize the promoting effect of the sentence-level intent information, a CRF module is used to focus on the slot value distribution at each position in the entire sentence. The feature matrix P is input into the CRF module to learn and extract the slot value distribution features at the sentence level. The CRF module includes a state transition matrix $D \in \mathbb{R}^{k \times k}$, where $D_{i,j}$ represents the probability of transitioning from state i to state j in a continuous sequence of time steps. Therefore, the CRF can use the slot value information from the past and future directions to predict the slot value at the current time step. For the output label sequence $S = \{S_1, S_2, \ldots, S_n\}$, the predicted score is calculated as follows:

$$\text{score}(S) = \sum_{i=1}^{n} P_{i,S_i} + \sum_{i=1}^{n+1} D_{S_{i-1},S_i} \tag{20}$$

Finally, the softmax function is used to obtain the optimal slot value extraction result $\bar{S}$.

$$\bar{S} = \arg\max_{S} \text{score}(S) \tag{21}$$

Training Method. During the training process of the bidirectional stack propagation network, the cross-entropy loss function is used to calculate the intent detection loss L_i and the slot filling loss L_s, and the goal is to maximize the intention y and the slot value sequence $\bar{S}$ log-likelihood probability:

$$\mathcal{L}_I = -\log p(y \mid X) \tag{22}$$

$$\mathcal{L}_{\bar{S}} = -\log p(\bar{S} \mid X) = -\sum_{i=1}^{n} \log p\left(\bar{S}_i \mid X\right) \tag{23}$$

To perform joint learning of intent detection and slot filling, the final loss function is the sum of the losses from both tasks:

$$\mathcal{L} = \mathcal{L}_I + \mathcal{L}_{\bar{S}} = -\log p(y \mid X) - \sum_{i=1}^{n} \log p\left(\bar{S}_i \mid X\right) \tag{24}$$

The Adam optimizer is used to optimize the model parameters and achieve fast convergence of the model. Adam is one of the commonly used optimizers in the field of deep learning. Its advantages include fast computation speed, automatic adjustment of learning rate, and good robustness for sparse gradients. Adam maintains an independent learning rate for each parameter and can automatically adjust the learning rate. At the beginning of training, the learning rate starts from zero and gradually increases, which helps to slow down the phenomenon of premature overfitting during the training process. As the training progresses, the learning rate gradually decreases and eventually converges to a small value, which helps to keep the model stable.

4   Experiments

4.1   Experimental Settings

To evaluate the efficiency of our proposed model, we conduct experiments on two benchmark datasets. One is the publicly ATIS dataset [10] containing audio recordings of flight reservations, and the other is the custom-intent-engines collected SNIPS dataset. Both datasets used in our paper follows the same format and partition as in [7].

To ensure the reliability of the experimental results, this chapter conducts 5 experiments for the same model on the same dataset and takes the average as the final result. The size of RNN hidden layer is set to 128, and the pre-trained model uses the Base version with a hidden layer size of 768. The Adam optimizer is used with an initial learning rate of 0.001, a batch size of 64, and a default training epoch of 20. During the training process, the early stop strategy is used. Every 1 epoch, the model's performance on the validation set is evaluated. When the performance on the validation set does not improve for 3 consecutive times, the training is stopped. In the performance evaluation on the test set, only the result of the best performance on the validation set for each model is taken.

4.2   Baselines

We compare our model with the existing baselines include:

- Joint Seq2Seq. proposed a multi-task modeling approach for jointly modeling intent detection and slot filling in a single recurrent neural network (RNN) architecture.
- Attention BiRNN. leveraged the attention mechanism to allow the network to learn the relationship between slot and intent.
- Slot-Gated. proposed the slot-gated joint model to explore the correlation of slot filling and intent detection better.
- Stack-Propagation. proposed a unidirectional stack propagation network, the output of intent recognition is directly used as the input for slot filling, and a word-level intent detection mechanism is used, which improves the performance of both intent detection and slot filling.
- S-P + BERT. Building on Stack-Propagation, using a pre-trained BERT model to encode the input sequence can effectively improve the performance of the model.
- JointBERT. leveraged BERT as the encoder for the input sequence and directly using BERT's output for slot filling and intent detection.

Table 1. Slot filling and intent detection results on two datasets.

Model	SNIPS			ATIS		
	Slot(F1)	Intent(Acc)	Overall(Acc)	Slot(F1)	Intent(Acc)	Overall(Acc)
Joint Seq2Seq	87.30	96.86	73.22	94.32	92.63	80.71
Attention BiRNN [11]	87.89	96.62	74.18	94.61	92.88	78.99
Slot-Gated [12]	88.83	97.06	75.57	95.32	94.14	82.60
Stack-Propagation [13]	94.21	98.03	86.94	95.94	96.95	86.52
S-P + BERT	97.06	**98.90**	92.81	96.08	97.79	88.58
JointBERT [14]	96.74	98.47	92.65	96.13	97.58	88.23
DABP	**97.41**	98.85	**93.02**	**96.90**	**97.93**	**89.50**

4.3   Overall Results

Following [7], we evaluate the NLU performance of slot filling using F1 score and the performance of intent detection using accuracy, and sentence-level semantic frame parsing using overall accuracy. The results are shown in Table 1.

From the table, we can see that our model archieves the state-of-the-art performance. In the SNIPS dataset, compared with the previous best model Stack-Propagation + BERT, we achieve 0.35% improvement in slot filling F1 score, reached almost the same level in intent detection accuracy, and 0.21% improvement in overall accuracy. In the ATIS dataset, we achieve 0.82% improvement in slot filling F1 score, 0.14% improvement in intent detection accuracy, and 0.92% improvement in overall accuracy. This indicated the effectiveness of our DABP model.

Building a bidirectional stack propagation network for slot filling and intent detection can further explore the semantic relationships between slots and intents. In the process of mutual influence, it can better promote the coordination and information sharing between the two tasks, thereby further improving the performance of the model. Additionally, domain augmented models can leverage the strong feature extraction and semantic modeling capabilities of pre-trained models, and for target domain task-oriented dialogue data that is lacking in the large-scale general domain corpora used for pre-training, they can extract semantic information for specific vocabulary or sentence patterns through masked language modeling tasks and part-of-speech tagging tasks, in order to enhance the pre-trained model for the target domain, and thus improve the model's performance on the target domain task.

4.4 Analysis

In Sect. 4.3, we have shown that our model achieves the state-of-the-art performance on the two datasets. In this section, we will analyze the reason of the improvement. First, we explore the effect of our proposed domain augmentation method. Finally, we study the effect of our proposed bidirectional stack propagation network.

Effect of Domain Augmentation Method. To verify the effectiveness of our domain augmentation method, we conduct experiments with the following ablations: no domain augmentation, only use the mask language model task and only use the part-of-speech tagging task.

Table 2 gives the result of the ablation experiment. The result show that masked language modeling task and part-of-speech tagging task have a certain effect on the performance improvement of pre-trained models. However, the effects of the two tasks are not linearly increasing, but have a certain interactive effect. That is, the effects of the two tasks complement each other, and the best augmentation effect is achieved by using both tasks for training. Further analysis shows that traditional pre-training methods using masked language tasks can only model the implicit information in semantics by learning the semantic representations of the user's sentence context, which leads to the model still being in an underfitting state. On the other hand, part-of-speech tagging tasks

Table 2. The domain augmentation ablation results on two datasets.

Model	SNIPS			ATIS		
	Slot(F1)	Intent(Acc)	Overall(Acc)	Slot(F1)	Intent(Acc)	Overall(Acc)
No Domain Augmentation	97.24	98.65	92.65	96.14	97.62	89.11
Only MLM Task	97.35	**98.87**	93.00	96.77	97.82	89.39
Only PoS-Tagging Task	97.29	98.68	92.87	96.40	97.69	89.26
DABP	**97.41**	98.85	**93.02**	**96.90**	**97.93**	**89.50**

can enhance language representations by learning the explicit information of the user's sentence's part-of-speech tagging, thereby further improving the pre-trained model's language understanding ability. Therefore, the joint training enhancement method our proposed can complement and promote each other, effectively improving the pre-trained model's language understanding ability and thus improving the understanding ability of task-oriented dialogue sentences in specific domains.

Effect of Bidirectional Stack Propagation Network. To verify the effectiveness of our bidirectional stack propagation network, we conduct experiments with the following ablations: no stack propagation, only use the slot-to-intent stack propagation and only use the intent-to-slot stack propagation.

Table 3. The bidirectional stack propagation ablation results on two datasets.

Model	SNIPS			ATIS		
	Slot(F1)	Intent(Acc)	Overall(Acc)	Slot(F1)	Intent(Acc)	Overall(Acc)
No Stack Propagation	97.09	98.66	92.79	96.19	97.97	89.07
Only Slot-to-Intent	97.24	98.73	**93.05**	96.34	97.89	89.11
Only Intent-to-Slot	97.22	98.70	92.89	96.75	97.75	89.44
DABP	**97.41**	**98.85**	93.02	**96.90**	**97.93**	**89.50**

Table 3 gives the result of the ablation experiment. The results show that both sub-networks of the bidirectional stack propagation network play a certain role. Removing one of the sub-networks will cause a decrease in the model's performance. Among them, removing the intent-to-slot stack propagation will cause the largest performance decrease in slot filling task, while removing the slot-to-intent stack propagation will cause the largest performance decrease in intent detection task. This also verifies the semantic correlation between slot filling and intent detection tasks in the model, as well as the effectiveness of information sharing between slot values and intents through mutual interaction during training iterations. Removing any sub-network will result in the loss of relevant feature information of the other sub-network, thereby affecting the model's performance.

5 Conclusion

In this paper, we propose a domain augmentation and bidirectional stack propagation model for natural language understanding. We use the MLM task and the proposed PoS-Tagging task to enhance PTMs with domain-specific knowledge include both implicit and explicit. Besides that, the proposed bidirectional stack-propagation mechanism propagates the information between the slot filling and intent detection tasks. Experiments on two datasets show the effectiveness of the proposed models and achieve the state-of-the-art performance. Besides, we explore and analyze the effect of each domain augmentation task and the effect of bidirectional stack propagation respectively.

Our approach leaves a lot of room for improvement. For example, we can further explore the task-oriented NLU for multi-turns dialogue. In addition, we can also explore the NLU based on explicit knowledge base, such as knowledge graphs. If the relationship information of these entities can be added to the model, it will help improve the expressiveness of the model. In the future, we will continue to explore the improvement of the proposed model in NLU.

Accknowledgement. This work is funded by Key-Area Research and Development Program of Guangdong Province, China (2021B0101190002).

References

1. Huang, M., Zhu, X., Gao, J.: Challenges in building intelligent open-domain dialog systems. ACM Trans. Info. Syst. **38**(3), 1–32 (2020)
2. Ren, H., Xu, W., Zhang, Y., et al.: Dialog state tracking using conditional random fields. In: Proceedings of the SIGDIAL 2013 Conference (2013)
3. Qian, C., Zhuo, Z., Wang, W.: Bert for joint intent classification and slot filling. arXiv preprint arXiv:1902.10909 (2019)
4. Brown, T., Mann, B., Ryder, N., et al.: Language models are few-shot learners. Adv. Neural. Inf. Process. Syst. **159**, 1877–1901 (2020)
5. Haihong, E., Niu, P., Chen, Z.. et al.: A novel bi-directional interrelated model for joint detection and slot filling. In: Proceedings of the 57th Annual Meeting of the Association for Computational Linguistics (2019)
6. Bing, L., Lane, I.: Attention-based recurrent neural network models for joint intent detection and slot filling. Interspeech 2016 (2016)
7. Libo, Q., Wanxiang, C., Li, Y., et al.: A stack-propagation framework with token-level intent detection for spoken language understanding. In: Proceedings of the 2019 Conference on Empirical Methods in Natural Language Processing and the 9th International Joint Conference on Natural Language Processing (2019)
8. Devlin, J., Chang, M., Lee, K., et al.: BERT: pre-training of deep bidirectional transformers for language understanding. In: Proceedings of the 2019 Conference of the North American Chapter of the Association for Computational Linguistics: Human Language Technologies, Volume 1 (Long and Short Papers) (2019)
9. Manning, C., Surdeanu, M., Bauer, J., et al.: The Stanford CoreNLP natural language processing toolkit. In: Proceedings of 52nd Annual Meeting of the Association for Computational Linguistics: System Demonstrations (2014)
10. Hemphill, C.T., Godfrey, J.J., Doddington, G.R.: The ATIS spoken language systems pilot corpus. In: Speech and Natural Language: Proceedings of a Workshop Held at Hidden Valley (1990)
11. Liu, B., Lane, I.: Attention-based recurrent neural network models for joint intent detection and slot filling. arXiv preprint arXiv:1609.01454 (2016)
12. Qin, L., Che, W., Li, Y., et al.: A stack-propagation framework with token-level intent detection for spoken language understanding. arXiv preprint arXiv:1909.02188 (2019)

13. Goo, C.W., Gao, G., Hsu, Y.K., et al.: Slot-gated modeling for joint slot filling and intent prediction. In: Proceedings of the 2018 Conference of the North American Chapter of the Association for Computational Linguistics: Human Language Technologies, vol. 2 (Short Papers), pp. 753–757 (2018)
14. Chen, Q., Zhuo, Z., Wang, W.: BERT for joint intent classification and slot filling. arXiv preprint arXiv:1902.10909 (2019)

Knowledge Graph Completion via Subgraph Topology Augmentation

Huafei Huang[1], Feng Ding[1], Fengyi Zhang[1], Yingbo Wang[1], Ciyuan Peng[2], Ahsan Shehzad[1], Qihang Lei[1], Lili Cong[3(✉)], and Shuo Yu[4]

[1] School of Software, Dalian University of Technology, Dalian 116620, China
hhuafei@outlook.com, dingfeng@dlut.edu.cn
[2] Institute of Innovation, Science and Sustainability, Federation University Australia, Ballarat, VIC 3353, Australia
[3] Department of Obstetrics and Gynecology, Affiliated Zhongshan Hospital of Dalian University, Dalian 116001, China
lili.cong.zsyy@outlook.com
[4] School of Computer Science and Technology, Dalian University of Technology, Dalian 116024, China
yushuo@dlut.edu.cn

Abstract. Knowledge graph completion (KGC) has achieved wide spread success as a key technique to ensure high-quality structured knowledge for downstream tasks (e.g., recommendation systems and question answering). However, within the two primary categories of KGC algorithms, the embedding-based methods lack interpretability and most of them only work in transductive settings, while the rule-based approaches sacrifice expressive power to ensure that the models are interpretable. To address these challenges, we propose KGC-STA, a knowledge graph completion method via subgraph topology augmentation. First, KGC-STA contains two topological augmentations for the enclosing subgraphs, including the missing relation completion for sparse nodes and the removal of redundant nodes. Therefore, the augmented subgraphs can provide more useful information. Then a message-passing layer for multi-relation is designed to efficiently aggregate and learn the surrounding information of nodes in the subgraph for triplet scoring. Experimental results in WN18RR and FB15k-237 show that KGC-STA outperforms other baselines and shows higher effectiveness.

Keywords: Knowledge Graph Completion · Graph Neural Network · Link Prediction · Graph Augmentation

1 Introduction

Knowledge Graphs (KGs) [14] is a sort of heterogeneous data structure, which contains a large number of facts that are represented by triplets (source entity, relation, target entity). Research based on KGs has attracted widespread interest and has been successfully applied in various fields such as Natural Language

F. Wu et al. (Eds.): SMP 2023, CCIS 1945, pp. 14–29, 2024.
https://doi.org/10.1007/978-981-99-7596-9_2

Processing [12], Recommender Systems [13,21], Question Answering [4,8], and Knowledge Reasoning [10]. However, while more knowledge graphs are created, the data explosion is accompanied by an increasingly serious data sparsity problem. Meanwhile, knowledge graphs often suffer from incomplete issues, and a data-rich and accurate knowledge graph requires continuous expert manpower to maintain, which is time-consuming and infeasible.

To alleviate the sparsity issue, the knowledge graph completion (KGC) technique, which can reduce the human cost in building and verifying knowledge graphs, has become a promising trend. KGC approaches focus on predicting potential entities or relations, thus making the knowledge graph complete. For example, in Fig. 1, a KGC method can infer that *K. De Bruyne* and *R. Mahrez* are teammates of *Haaland* by observing that *Haaland, K. De Bruyne*, and *R. Mahrez* are all part of the *Man. City* football club, and then add links to them.

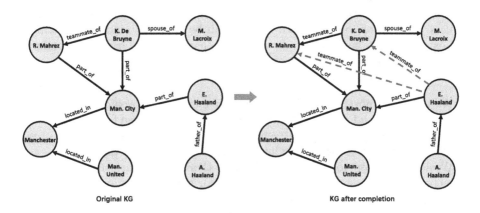

Fig. 1. Illustration of knowledge graph completion.

Knowledge graph completion has become a hot topic, and the approaches are mainly divided into two branches, i.e., embedding-based and rule-based methods. Embedding-based approaches aim to map entities and relations in the knowledge graph to a low-dimensional space, then predict the missing parts of triplets. Classical knowledge embedding approaches are translation models, scoring triplets' correctness based on the distance between entities, such as TransE [2], TransD [7], and TransG [23]. More recently, deep learning approaches [1,5] learn more expressive knowledge embedding in an end-to-end manner. Graph Neural Networks, which can be applied directly to non-Euclidean data, provide an effective and feasible direction for knowledge graph completion. Vashishth *et al.* [20] introduced graph convolutional networks to aggregate multiple relational information in knowledge graphs to achieve more accurate embeddings of entities and relations. Teru *et al.* [17] augment the contextual information of triplets by extracting subgraphs to evaluate facts more effectively. Despite considerable progress, these methods lack interpretability, and most of them are only under transductive settings.

Another branch, rule-based methods, instead of using a black box way, focuses on explicitly learning logical rules to complete the knowledge graph. Yang *et al.* [25] proposed a differentiable approach to learn rules end-to-end and then utilize them for knowledge graph completion when inference. DRUM [15] can mine rules for variable lengths and is equipped with an RNN model to improve learning ability. However, interpretability is obtained at the expense of losing expressive power in the rule-based approaches.

To remedy these defects, in this paper, we propose a **K**nowledge **G**raph **C**ompletion method via **S**ubgraph **T**opology **A**ugmentation dubbed KGC-STA. Firstly, an enclosing subgraph is extracted for each triplet in the knowledge graph, and then we optimize the subgraph topology, including adding relations for sparse nodes and dropping redundant nodes. Then we design a message-passing layer for multi-relation to efficiently aggregate and learn the surrounding information of nodes in the subgraph. Finally, we can accurately predict the relations between entities after optimization by reducing a loss based on scoring triplets. In addition, our proposed method can be applied to inductive settings and extended to large-scale and distributed scenarios in the future. We conduct extensive experiments and verify the effectiveness of our model.

Our contributions are summarized as follows:

- We propose a novel method KGC-STA for knowledge graph completion. By considering the information of enclosing subgraphs extracted from triplets and using topology augmentation techniques, KGC-STA shows superior expressiveness while ensuring certain interpretability in the KGC task.
- We design two topological augmentation methods for the enclosing subgraphs, including the missing relation completion for sparse nodes and the removal of redundant nodes. Thus, it is ensured that the optimized subgraphs can provide more useful information in the KGC process.
- We compare KGC-STA with ten popular KGC methods on two real-world datasets, WN18RR and FB15k-237. The experimental results show that KGC-STA outperforms other algorithms, thus proving the effectiveness and learning ability of our proposed KGC-STA.

The remainder of the paper is organized as follows: In Sect. 2, we illustrate the related works. Section 3 give the necessary concepts and preliminaries. Section 4 introduces the proposed KGC-STA framework. In Sect. 5, we conduct sufficient experiments on our approach and analyze the results in detail. Section 6 summarizes this work.

2 Related Works

Knowledge graph completion is one of the significant applications in knowledge graph representation learning and has attracted the interest of many researchers. Various approaches have been proposed in the current works, and we introduce embedding-based and rule-based approaches.

2.1 Embedding-Based Methods

Embedding-based approaches aim to project entities in the knowledge graph to a low-dimensional space and enforce the entities to satisfy certain positional relations in the space. TransE [2] models relations by viewing them as translating operators and then computing the entities. This idea has inspired a series of subsequent works [7,9,22]. ComplEx [19] introduces complex embedding space, allowing the method to capture symmetric and antisymmetric relations. Dettmers *et al.* [5] incorporate 2D convolutional networks into knowledge graph representation learning and then extract effective features to predict missing information in triplets using fewer training parameters. HypER [1] is a tensor factorization-based model for relation prediction. It enhances prediction tasks by converting relations into convolutional kernels and reducing parameters simultaneously. RotatE [16] views each relation as a rotation from the source entity to the target entity in a complex space. Thus, it can capture inversion, composition symmetry, and antisymmetry patterns. Vashishth *et al.* [20] introduce graph convolutional networks to implement knowledge graph embedding, which updates the entity embeddings by three different operations to learn efficient representations. Teru *et al.* [17] proposed GraiL, an inductive relation prediction model based on graph neural networks.

2.2 Rule-Based Methods

Rule-based methods are highly interpretable and can be applied to knowledge graph completion in inductive scenarios. Neural LP [25] is the first method to learn first-order logic rules in an end-to-end and differentiable manner. This approach has been successfully applied to knowledge base reasoning tasks. Meilicke *et al.* [11] designed a rule mining algorithm named RuleN, which is able to estimate confidence based on a set of randomly selected samples. Sadeghian *et al.* [15] proposed an end-to-end rule learning method, DRUM, which enables the model to learn fixed length rules as well as variable length rules while reducing the number of learnable parameters and enhancing interpretability. In addition, the authors introduced LSTM to ensure the learning capability of DRUM.

3 Preliminaries

3.1 Definitions

We give several necessary definitions before introducing our work.

1 *Knowledge Graph*: *A knowledge graph can be represented as* $G = \{\mathcal{E}, \mathcal{R}, \mathcal{TP}\}$*, where* $\mathcal{E}$ *denotes entity set,* $\mathcal{R}$ *is the relation set, and* $\mathcal{TP} = \{(h, r, t)\}$ *is a set of fact triplets.*

Then we introduce the definition of network motif as follows.

2 Network Motif: *Network motifs are a sort of low-order structure sub-graphs that repeat themselves within a particular network or among a collection of networks.*

We then introduce Entity Motif Degree (EMD), an intuitive and effective measurement of the structural information of nodes in a graph. This concept was pioneered by Yu *et al.* [27] and has been widely used in graph learning applications [24,26,28].

3 Entity Motif Degree: *The Entity Motif Degree (EMD) is used to count the number of motifs contained in a given node in the graph. Given a knowledge graph $G = \{\mathcal{E}, \mathcal{R}, \mathcal{TP}\}$, the EMD of the entity i is $EMD_M(i)$, which represents the number of instance regarding motif type M containing entity i.*

Moreover, to better introduce the reader to these definitions, we give an illustration of motifs and Entity Motif Degree in Fig. 2, where motif types M31, M32, M41, M42, M43 are used in this paper.

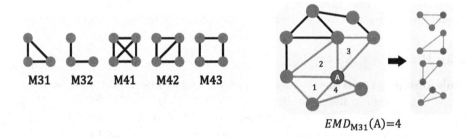

$$EMD_{M31}(A) = 4$$

Fig. 2. An illustration of motifs and Entity Motif Megree (EMD).

To facilitate the subgraph extraction against triplets in knowledge graphs, we introduce the enclosing subgraph [29].

4 Enclosing Subgraph: *For a given graph G, the k-hop enclosing subgraph of the two nodes u and v, denoted as G_{uv}^k, is an induced subgraph consisting of the intersection of the k-hop neighbors $\mathcal{N}_k(u)$ and $\mathcal{N}_k(u)$.*

3.2   Problem Formulation

Given a knowledge graph $G = \{\mathcal{E}, \mathcal{R}, \mathcal{TP}\}$, knowledge graph completion (KGC) is the task of predicting and replenishing the missing parts of triples. This paper focuses on the link prediction task and applies our approach in an inductive setting.

4   Methodology

The proposed KGC-STA is comprised of three core parts, which are **Subgraph Extraction and Rule-Based Augmentation**, **Subgraph Pruning via Entity Motif Degree**, and **Completion Scoring and Optimization**, respectively. The whole framework of KGC-STA is outlined in Fig. 3.

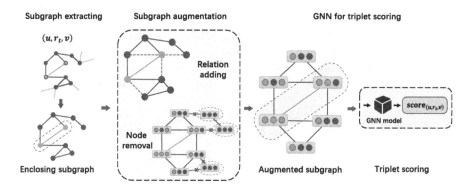

Fig. 3. The framework of KGC-STA.

4.1   Subgraph Extraction and Rule-Based Augmentation

We first extract enclosing subgraphs for all triplets in the knowledge graph. For a given triplet (u, r_t, v), we extract the enclosing subgraph $G_{(u,r_t,v)}$ according to the target nodes u, v, and target relation r_t. We believe that the surrounding structure around the node contains rich contextual information. Therefore, extracting and using this information will help improve the model performance in knowledge graph inference. In addition, extracting subgraphs for each triplet can smoothly convert link predictions to subgraph-level tasks, i.e., graph classification.

However, the subgraph extraction may fail when faced with sparse entity nodes, and such nodes lacking context information tend to be ignored in subgraph construction and message passing. To alleviate this drawback, we introduce rule mining methods to enhance the connectivity of sparse nodes. Here we first evaluate the sparsity of an entity u.

$$\text{sparsity}(u) = 1 - \frac{\text{freq}(u) - \text{freq}_{\text{min}}}{\text{freq}_{\text{max}} - \text{freq}_{\text{min}}}, \tag{1}$$

where $\text{freq}(u)$ is the frequency of u occurrence as a head or tail entity from triplet in the knowledge graph dataset, and freq_{min} and freq_{max} are the minimum and maximum frequencies, respectively. A larger $\text{sparsity}(u)$ indicates a higher sparsity of entity u. Conversely, when $\text{sparsity}(u)$ is close to 0, entity u is connected to rich and sufficient neighborhoods. Here we set the threshold θ and classify u as a sparse entity when sparsity$(u) < \theta$. Then we perform structural completion on entities by the rule-based method.

After that, we infer relations in the extracted subgraph for the determined sparse entity u. Specifically, we connect the potential triplet $(u, ?, v^*)$, where v^* is not the direct neighborhood node of the target sparse entity u in the enclosing subgraph. To reduce the computation, we conduct completion for the target nodes u and v in a triplet (an enclosing subgraph contains only two) and finally form a new subgraph G'_{uv}. Completion relies on rules obtained by learning; specifically, we apply differentiable logic rule learning methods [11,15,25] in the knowledge graph beforehand to obtain rules and perform relation completion on target nodes after enclosing subgraphs extracted. For example, to complete the instance $(u, ?, v^*)$ to (u, r_n, v^*), the new relation r_n is obtained as follows.

$$r_n(X, Y) \leftarrow r_1\left(X, Z_1\right) \wedge r_2\left(Z_1, Z_2\right) \wedge r_3\left(Z_2, Z_3\right) \ldots \wedge r_k\left(Z_k, Y\right), \qquad (2)$$

where X and Y are variables of entities, $r_1, r_2, \ldots, r_k$ are all binary relations. Thus we can detect potential relations through the learned rules (dubbed path) between X and Y, thus adding new links to two entities.

4.2   Subgraph Pruning via Entity Motif Degree

The subgraph completed via rule acquires rich structural information but simultaneously brings a huge unnecessary redundancy. This makes it hard to get distinguishable embeddings through the current structure, so we improve it from both structure and attribute perspectives.

We use the Entity Motif Megree (EMD) to measure the redundancy of nodes in the subgraph. Here we calculate the EMD of several motifs of all nodes in the knowledge graph in advance and then use them to determine whether the nodes are redundant. Specifically, for a given motif M, we obtain the EMD of nodes in the subgraph (except the target node) and compare them with the global average EMD. Once the EMD of node u is less than the average for all types of motifs, we remove it from the enclosing subgraph. This is based on the assumption that nodes with a higher EMD (1) contain richer structures and (2) have more structural information completed by rule mining. Given a subgraph G_{sub}, the EMD matrix $\mathbf{M}_d \in \mathbb{R}^{N_{sub} \times N_M}$ is formed as follows.

$$\mathbf{M}_d = \left[\overrightarrow{E}_{md}(u_1), \overrightarrow{E}_{md}(u_2), \overrightarrow{E}_{md}(u_3), \ldots, \overrightarrow{E}_{md}(u_{N_{sub}})\right]^T, \qquad (3)$$

where $\overrightarrow{E}_{md}(u_i) \in \mathbb{R}^{N_M}$ is the EMD vector of i-th node (entity) in G_{sub}, N_{sub} is the subgraph's number of node, and N_M denotes the number of motif types used in this work (i.e. M31, M32, M41, M42, M43). After the node pruning, we obtain the augmented extracted enclosing subgraph for each triplet in the knowledge graph.

Entities in the knowledge graph do not have initial attribute information. To solve the issue, we compute encoding structure information for nodes in the subgraph as attributes. In particular, given a enclosing subgraph G_{uv}, for any node i, we first compute the shortest distances $d(i, u)$ and $d(i, v)$ from i to the two target nodes u and v, respectively, and then concatenate them as the initialized attribute $\mathbf{x}_i = [d(i, u), d(i, v)]$ of i. Especially for two root nodes u and v, the attributes are obviously $[0, 1]$ and $[1, 0]$.

4.3   Completion Scoring and Optimization

We complete the knowledge graph in a subgraph-level manner by extracting closed subgraphs. For an extracted subgraph G_{uv}, we can apply the Graph Neural Network (GNN) approach to compute the triplet score of a potential truth (u, r_t, v) because the message-passing mechanism [6] in GNNs can easily meet the requirements. For given node i, we aggregate the neighborhood nodes based on different relationships in the subgraph shown in Eq. (4).

$$
\mathbf{a}_t^l = \varphi \left(\sum_{re \in \mathcal{R}} \sum_{j \in \mathcal{N}_i^{re}} \alpha_{ij}^l \mathbf{h}_j^{(l)} \mathbf{W}_{re}^{(l)} + \mathbf{h}_i^{(l)} \mathbf{W}_0^{(l)} \right),
\tag{4}
$$

where $\alpha_{ij}^l \in \mathbb{R}$ is the coefficient at l-th layer for the relation re, $\mathbf{W}_{re}^{(l)}$ and $\mathbf{W}_0^{(l)}$ are weight matrixes for relation and the node itself at l-th layer, respectively, $\varphi(\cdot)$ is the aggregator in message-passing.

The α_{ij}^l value can be learned directly or more complexly. In this work, we utilize the attention mechanism by multi-layer perceptron to learn the equation in the following way: α_{ij}^l adaptively.

$$
\alpha_{ij}^l = \text{MLP} \left([\mathbf{h}_i^{l-1} \oplus \mathbf{h}_j^{l-1} \oplus \mathbf{r}_{re} \oplus \mathbf{r}_{triplet}] \right),
\tag{5}
$$

where MLP$(\cdot)$ is a two-layer method and transforms dimension to one to get the coefficient value, $\oplus$ means concatenate operation to combine different representation, the $\mathbf{h}_i^{l-1}$ and $\mathbf{h}_j^{l-1}$ are embeddings of self and neighborhood nodes, the $\mathbf{r}_{re}$ and $\mathbf{r}_{triplet}$ are relation representations for the aggregation and target triplet (which induce the subgraph).

After aggregation in Eq. (4), we employ a linear transformation plus the ReLU$(\cdot)$ function to complete the output of the l-th layer GNN and use it as the input for the next layer.

$$
\mathbf{h}_t^l = \text{ReLU} \left(\mathbf{a}_t^l \mathbf{W}_{\text{self}}^l + \mathbf{b}_t^l \right).
\tag{6}
$$

We get the embeddings at the node level by running the message passing layer by layer, and then we try to get the subgraph representation. Here we choose MEAN as the READOUT$(\cdot)$ function, shown as $\mathbf{h}_{G_{(u, r_t, v)}} = \frac{1}{|\mathcal{V}_{G_{(u, r_t, v)}}|} \sum_{i \in \mathcal{V}} \mathbf{h}_i^L$. To effectively score a given triplet (u, r_t, v), we consider multi-level information obtained by KGC-STA, i.e., combining the subgraph, target nodes, and target relation representations, and then the score is obtained via projection.

$$\text{score}_{(u,r_t,v)} = \left[\mathbf{h}_{G(u,r_t,v)} \oplus \mathbf{h}_u^L \oplus \mathbf{h}_v^L \oplus \mathbf{r}_t\right] \mathbf{W}_s. \tag{7}$$

We select Max-margin Loss to evaluate the state of our proposed KGC-STA until optimization converges. In the implementation, we generate negative samples for the triplet in datasets. The detailed loss is as follows.

$$\mathcal{L} = \sum_{i=1}^{|\mathcal{E}|} \max\left(0, \text{score}_{neg(u,r_t,v)} - \text{score}_{pos(u,r_t,v)} + \gamma\right), \tag{8}$$

where $neg(\cdot)$ and $pos(\cdot)$ are negative and positive samples, respectively. $\gamma \in \mathbb{R}$ denotes the margin in the loss.

5   Experiments

In this section, we perform experiments for the proposed KGC-STA model, including link prediction, ablation study, and parameter sensitivity analysis.

5.1   Experiment Settings

Datasets. We use benchmark datasets from real-world scenarios in our experiments, including WN18RR [5] and FB15k-237 [18]. For WN18RR, there are 93,003 triplets in the dataset, with 40,943 entities and 11 types of relations. FB15k-237 is a subset created from FB15k [2], processed by dropping redundant relations and inversed triples. FB15k-237 contains 310,079 triples, 14,505 entities, and 237 types of relations. We sampled four sets of training and testing graphs from the original datasets and denoted them as v1, v2, v3, and v4 in the experiments. The strategy for dataset splitting is referenced from the work of Teru *et al.* [17]. During the experiments, all models (ours and baselines) are trained on the training graphs and tested on the test graphs, and the test ratio is set to 10%. Table 1 shows the detailed statistical information of WN18RR and FB15k-237.

Baselines. We have 10 baselines in our experiments and divide these methods into two categories referring to Sect. 2.
  (1) Embedding-based method:

- **TransE** [2]: TransE is a classical knowledge graph embedding learning approach. TransE treats relations as translations from head entities to tail entities and gives score functions based on Euclidean distances to optimize the objectives.
- **ComplEx** [19]: ComplEx is a bilinear model that introduces complex vectors and can effectively capture symmetric and antisymmetric relations.

Table 1. Statistical information of WN18RR and FB15k-237.

		WN18RR			FB15K-237		
		Relations	Nodes	Links	Relations	Nodes	Links
v1	#train	9	2,746	6,678	183	2,000	5,226
	#test	9	922	1,191	146	1,500	2,404
v2	#train	10	6,954	18,968	203	3,000	12,085
	#test	10	2,923	4,863	176	2,000	5,092
v3	#train	11	12,078	32,150	218	4,000	22,394
	#test	11	5,084	7,470	187	3,000	9,137
v4	#train	9	3,861	9,842	222	5,000	33,916
	#test	9	7,208	15,157	204	3,500	14,554

- **ConvE** [5]: ConvE incorporates convolutional neural networks into knowledge graph embedding learning, applying 2D convolution to extract triplet features to infer missing information.
- **HypER** [1]: HypER is a relation prediction model using tensor factorization that regards relations as convolutional kernels, reducing training parameters and improving the effectiveness of downstream tasks.
- **RotatE** [16]: RotatE maps entities into complex space to model various complex relations such as symmetry/antisymmetry, inversion, composition, etc., improving the performance in link prediction.
- **COMPGCN** [20]: COMPGCN focuses on multiple relations information and applies graph convolution networks to improve entity and relation embeddings.
- **GraIL** [17]: GraIL is a GNN-based framework for inductive knowledge graph reasoning, which can extract local subgraph information to learn entity-independent relational semantics.

(2) Rule-based method:

- **Neural LP** [25]: Neural LP is an end-to-end differentiable neural logic framework that efficiently learns rules and applies them to knowledge inference tasks.
- **DRUM** [15]: RUM is an inductive approach for relation prediction that learns and mines first-order logic rules differently, thus conducting completion on unseen knowledge graphs.
- **RuleN** [11]: RuleN explicitly extracts the information from path rules and uses an ensemble to achieve better results.

Evaluation. We apply AUC-PR, MRR, and Hits@k evaluation metrics throughout our experiments. (1) **AUC-PR** is a metric for classification tasks, which calculates the area under the Precision-Recall curve (P-R Curve). A Steeper P-R curve indicates better model performance. Therefore, the ideal P-R curve value is 1, while the worst one is 0.5. (2) **MRR** (Mean Reciprocal Rank) is used for evaluating ranking algorithms. The performance of a system is measured by ranking the correct search result value in all answers to queries. The equation of MRR is $MRR = \frac{1}{Q} \sum_{i=1}^{|Q|} \frac{1}{\text{rank}_i}$, where $|Q|$ is the number of search results, rank_i refers to the position of the highest-ranked answer result. (3) **Hits@k** computes the proportion of triplets whose ranking lies in the top-k in link prediction, and a larger value of Hits@k represents better prediction performance of the model. The equation is formulated as $Hits@k = \frac{1}{|S|} \sum_{i=1}^{|S|} \Pi(\text{rank}_i \leq k)$, where $\Pi(\cdot)$ is the indicator function, that is, when the condition is true then the function value is 1, otherwise 0.

Implementation Details. All experiments in this work are implemented in Python 3.8 on the Ubuntu operation system. Moreover, we use the framework Pytorch 1.8.1 and the DGL Library to design and conduct experiments. The learning rate is set to 0.01, and the running epoch is set to 100. Other hyperparameters followed the settings from Teru *et al.* [17].

5.2 Link Prediction Results

We first conduct link prediction experiments for all methods on the benchmark datasets WN18RR and FB15k-237 and then give the results as shown in Table 2. The symbol '*' from the table indicates that this approach's experimental result is taken from the original paper. To facilitate comparative observation, we highlight the best results of the performance in bold and the second best in underlined. The symbol '-' indicates no corresponding experiment in the original paper.

Table 2. Experimental results on WN18RR and FB15k-237.

Method	WN18RR				FB15k-237			
	MRR	Hits@10	Hits@3	Hits@1	MRR	Hits@10	Hits@3	Hits@1
TransE*	0.226	0.501	-	-	0.294	0.465	–	–
ComplEx*	0.44	0.51	0.44	0.39	0.247	0.428	0.275	0.158
ConvE*	0.43	0.52	0.44	0.4	0.325	0.501	0.356	0.237
HypER*	0.465	0.522	0.477	0.436	0.341	0.52	0.376	0.252
RotatE*	0.476	<u>0.571</u>	0.492	0.428	0.338	0.533	0.375	0.241
COMPGCN*	<u>0.479</u>	0.546	<u>0.494</u>	<u>0.443</u>	<u>0.355</u>	<u>0.535</u>	<u>0.394</u>	<u>0.264</u>
Neural LP*	0.435	0.566	0.434	0.371	0.24	0.362	–	–
DRUM*	0.435	0.568	0.435	0.37	0.25	0.373	0.271	0.187
KGC-STA	**0.484**	**0.657**	**0.516**	**0.475**	**0.361**	**0.557**	**0.416**	**0.274**

As shown in the table, in general, KGC-STA has the best results compared to all baselines in both datasets, with significant advantages in all evaluation metrics. From the result in WN18RR, KGC-STA improves 1.04% over the second-best COMPGCN in MRR and significantly gains 15.06%, 4.45%, and 7.22% over the second performance in Hits@10, Hits@3, and Hits@1 metrics, respectively. Similarly, in FB15k-237, KGC-STA shows significant outperformance in both MRR and Hits@10, Hits@3, and Hits@1, reflecting the effectiveness of the proposed KGC-STA. Moreover, there are structural sparsity issues in WN18RR, where the average number of links to one entity is only 2.1, while the value on dataset FB15k-237 is 18.7. This reflects that KGC-STA can also be effectively applied to completion scenarios under sparse knowledge graphs, where rule-based augmentation compensates for embedding sparse nodes.

Next, we conduct inductive link prediction experiments for KGC-STA. We select Neural LP, DRUM, RuleN, and GraIL as the comparison methods and performed on two datasets (already processed), WN18RR and FB15k-237, respectively. The experimental results are shown in Tables 3 and 4, respectively.

Table 3. Experimental results on WN18RR.

Method	AUC-PR				Hits@10			
	v1	v2	v3	v4	v1	v2	v3	v4
Neural LP	86.02	83.78	62.9	82.06	74.37	68.93	46.18	67.13
DRUM	86.02	84.05	63.2	82.06	74.37	68.93	46.18	67.13
RuleN	90.26	89.01	76.46	85.75	80.85	78.23	53.39	71.59
GraIL	_94.32_	_94.18_	_85.8_	_92.72_	_82.45_	_78.68_	_58.43_	_73.41_
KGC-STA	**94.41**	**97.24**	**86.96**	**95.06**	**84.04**	**81.63**	**62.81**	**76.35**

The results presented in WN18RR show that KGC-STA also outperforms other methods in the inductive scenario. This is attributed to the fact that KGC-STA selects high-quality target subgraphs and applies a relatively lightweight likelihood scoring method to ensure excellent results. Especially in the Hits@10 metric, KGC-STA outperforms the suboptimal results by 1.93%, 3.75%, 5.51%, and 3.47% on v1, v2, v3, and v4 parts, respectively.

Table 4. Experimental results on FB15k-237.

Method	AUC-PR				Hits@10			
	v1	v2	v3	v4	v1	v2	v3	v4
Neural LP	69.64	76.55	73.95	75.74	52.92	58.94	52.9	55.88
DRUM	69.71	76.44	74.03	76.2	52.92	58.73	52.9	55.88
RuleN	75.24	88.7	91.24	91.79	49.76	77.82	**87.69**	85.6
GraIL	**84.69**	**90.57**	**91.68**	**94.46**	64.15	**81.8**	82.83	89.29
KGC-STA	_82.31_	_90.23_	_91.64_	_94.19_	**65.61**	_81.49_	_83.16_	**89.74**

As for the results on FB15k-237, KGC-STA still outperforms most other algorithms against a non-sparse relations dataset, with only slightly lags behind the best method, GraIL. We believe this is because the relations in FB15k-237 are enough to support all baselines to learn sufficiently during training and output high-quality knowledge representation. This also means that the rule-based graph augmentation in KGC-STA does not achieve much performance gain in FB15k-237.

5.3 Ablation Study

To explore how each part of the KGC-STA framework affects the experimental performance, we conduct an ablation analysis on three core parts in KGC-STA to determine their effectiveness. Specifically, we remove/replace each of the three modules of the KGC-STA algorithm respectively, namely the rule-based subgraph augmentation, the attention mechanism in graph neural networks, and the graph topology pruning, and then conduct inductive link prediction. The results (Hits@10) of the ablation study on the two datasets (v1 part) are shown in Table 5.

Table 5. Ablation test results (Hits@10).

	FB15k-237(v1)	WN18RR(v1)
KGC-STA	**65.61**	**84.04**
KGC-STA w/o rule augmentation	65.12	80.51
KGC-STA w/o graph attention	63.17	82.04
KGC-STA w/o subgraph pruning	60.74	80.36

The "**w/o rule augmentation**" in the table means KGC-STA extracts all k-hop neighbors of the target node to construct subgraphs without using the rule-based subgraph augmentation strategy. Here we know the necessity to use rule-based extraction instead of simply considering the subgraphs induced by all k-hop neighbors of the target node. In such settings, we observe a sharp degradation performance, while removing the subgraph extraction module would cause overfitting and make the AUC-PR exceed 0.99 during training. The "**w/o graph attention**" setting removes the attention part in KGC-STA , i.e., makes every α_{ij}^{l} in Eq. (5) a fixed and equal value in the model, thus eliminating the role of the attention mechanism. It can be seen that the performance of KGC-STA without the attention mechanism is significantly poor. In "**w/o subgraph pruning**", we replace the subgraph pruning in this work with the subgraph sampling in RLvRL [11]. The results show degraded performance and the experiment runs longer.

5.4   Parameter Analysis

In this section, we first investigate the impact of different aggregators on the performance of KGC-STA during message passing in GNN. We use three aggregation ways in Eq. (4), and the results are shown in Table 6. Specifically, 'SUM' means summation aggregator φ, 'MLP' represent that features are aggregated by multi-layer perceptron, and 'GRU' is a kind of the Recurrent Neural Network (RNN) [3].

Table 6. The impact of message aggregation strategies on prediction performance.

	AUC-PR		Hits@10	
Method	WN18RR(v1)	FB15k-237(v1)	WN18RR(v1)	FB15k-237(v1)
SUM	**94.41**	**82.31**	84.04	**65.61**
MLP	94.20	79.34	84.04	61.71
GRU	91.09	78.93	84.04	57.80

In general, it can be seen that when choosing the 'SUM' aggregator, KGC-STA achieves the overall best results on both datasets. In particular, on the FB15k-237 dataset, the experimental results of the SUM aggregator are significantly superior. Although there are only marginal gains on the WN18RR, we think this is partially attributed to the small number of relations on the dataset.

Then we test the effect of different neighbor hop numbers on the experimental results of link prediction when extracting subgraphs for target nodes. The results are shown in Table 7. As can be seen, the result suggests that when the hop number is set to 2, KGC-STA could outperform other versions.

Table 7. Effect of extracting subgraph hops on prediction performance.

	AUC-PR		Hits@10	
Method	WN18RR(v1)	FB15k-237(v1)	WN18RR(v1)	FB15k-237(v1)
hop=1	92.43	71.89	84.04	51.95
hop=2	**95.11**	**83.02**	84.04	**59.02**
hop=3	94.41	82.31	84.04	57.80

6   Conclusion

Ensuring the quality and completeness of the knowledge graphs has become an important task. In this paper, we propose a knowledge graph completion method KGC-STA, which can effectively conduct topological augmentation for enclosing subgraphs and use a multi-relation graph neural network to aggregate structural information. Our KGC-STA can be applied to inductive settings easily while improving the quality and interpretability of KGC. The experimental results prove that our KGC-STA performs better than other popular approaches.

References

1. Balažević, I., Allen, C., Hospedales, T.M.: Hypernetwork knowledge graph embeddings. In: International Conference on Artificial Neural Networks, pp. 553–565 (2019)
2. Bordes, A., Usunier, N., García-Durán, A., Weston, J., Yakhnenko, O.: Translating embeddings for modeling multi-relational data. In: Advances in Neural Information Processing Systems, pp. 2787–2795 (2013)
3. Chung, J., Gulcehre, C., Cho, K., Bengio, Y.: Empirical evaluation of gated recurrent neural networks on sequence modeling. arXiv preprint arXiv:1412.3555 (2014)
4. Das, R., et al.: Knowledge base question answering by case-based reasoning over subgraphs. In: International Conference on Machine Learning, pp. 4777–4793 (2022)
5. Dettmers, T., Minervini, P., Stenetorp, P., Riedel, S.: Convolutional 2D knowledge graph embeddings. In: Proceedings of the AAAI Conference on Artificial Intelligence (2018)
6. Gilmer, J., Schoenholz, S.S., Riley, P.F., Vinyals, O., Dahl, G.E.: Neural message passing for quantum chemistry. In: International Conference on Machine Learning, pp. 1263–1272 (2017)
7. Ji, G., He, S., Xu, L., Liu, K., Zhao, J.: Knowledge graph embedding via dynamic mapping matrix. In: Proceedings of the Annual Meeting of the Association for Computational Linguistics, pp. 687–696 (2015)
8. Jiang, L., Usbeck, R.: Knowledge graph question answering datasets and their generalizability: are they enough for future research? In: Proceedings of the International ACM SIGIR Conference on Research and Development in Information Retrieval, pp. 3209–3218 (2022)
9. Lin, Y., Liu, Z., Sun, M., Liu, Y., Zhu, X.: Learning entity and relation embeddings for knowledge graph completion. In: Proceedings of the AAAI Conference on Artificial Intelligence (2015)
10. Mavromatis, C., et al.: TempoQR: temporal question reasoning over knowledge graphs. In: Proceedings of the AAAI Conference on Artificial Intelligence, pp. 5825–5833 (2022)
11. Meilicke, C., Fink, M., Wang, Y., Ruffinelli, D., Gemulla, R., Stuckenschmidt, H.: Fine-grained evaluation of rule-and embedding-based systems for knowledge graph completion. In: International Semantic Web Conference, pp. 3–20 (2018)
12. Melluso, N., Grangel-González, I., Fantoni, G.: Enhancing industry 4.0 standards interoperability via knowledge graphs with natural language processing. Comput. Ind. **140**, 103676 (2022)
13. Park, S.J., Chae, D.K., Bae, H.K., Park, S., Kim, S.W.: Reinforcement learning over sentiment-augmented knowledge graphs towards accurate and explainable recommendation. In: Proceedings of the ACM International Conference on Web Search and Data Mining, pp. 784–793 (2022)
14. Peng, C., Xia, F., Naseriparsa, M., Osborne, F.: Knowledge graphs: opportunities and challenges. Artificial Intelligence Review, pp. 1–32 (2023)

15. Sadeghian, A., Armandpour, M., Ding, P., Wang, D.Z.: DRUM: end-to-end differentiable rule mining on knowledge graphs. In: Advances in Neural Information Processing Systems, pp. 15321–15331 (2019)
16. Sun, Z., Deng, Z.H., Nie, J.Y., Tang, J.: RotatE: knowledge graph embedding by relational rotation in complex space. In: International Conference on Learning Representations (2019)
17. Teru, K., Denis, E., Hamilton, W.: Inductive relation prediction by subgraph reasoning. In: International Conference on Machine Learning, pp. 9448–9457 (2020)
18. Toutanova, K., Chen, D., Pantel, P., Poon, H., Choudhury, P., Gamon, M.: Representing text for joint embedding of text and knowledge bases. In: Proceedings of the Conference on Empirical Methods in Natural Language Processing, pp. 1499–1509 (2015)
19. Trouillon, T., Welbl, J., Riedel, S., Gaussier, É., Bouchard, G.: Complex embeddings for simple link prediction. In: International Conference on Machine Learning, pp. 2071–2080 (2016)
20. Vashishth, S., Sanyal, S., Nitin, V., Talukdar, P.: Composition-based multirelational graph convolutional networks (2020)
21. Wang, X., Liu, K., Wang, D., Wu, L., Fu, Y., Xie, X.: Multi-level recommendation reasoning over knowledge graphs with reinforcement learning. In: Proceedings of the ACM Web Conference, pp. 2098–2108 (2022)
22. Wang, Z., Zhang, J., Feng, J., Chen, Z.: Knowledge graph embedding by translating on hyperplanes. In: Proceedings of the AAAI Conference on Artificial Intelligence (2014)
23. Xiao, H., Huang, M., Zhu, X.: TransG: a generative model for knowledge graph embedding. In: Proceedings of the Annual Meeting of the Association for Computational Linguistics, pp. 2316–2325 (2016)
24. Xu, J., et al.: Multivariate relations aggregation learning in social networks. In: Proceedings of the ACM/IEEE Joint Conference on Digital Libraries in 2020, pp. 77–86 (2020)
25. Yang, F., Yang, Z., Cohen, W.W.: Differentiable learning of logical rules for knowledge base reasoning. In: Advances in Neural Information Processing Systems. pp. 2319–2328 (2017)
26. Yu, S., Xia, F., Wang, Y., Li, S., Febrinanto, F.G., Chetty, M.: PANDORA: deep graph learning based COVID-19 infection risk level forecasting. IEEE Transactions on Computational Social Systems, pp. 1–14 (2022)
27. Yu, S., Xia, F., Xu, J., Chen, Z., Lee, I.: OFFER: a motif dimensional framework for network representation learning. In: Proceedings of the ACM International Conference on Information and Knowledge Management, pp. 3349–3352 (2020)
28. Yuan, X., Zhou, N., Yu, S., Huang, H., Chen, Z., Xia, F.: Higher-order structure based anomaly detection on attributed networks. In: IEEE International Conference on Big Data (Big Data), pp. 2691–2700 (2021)
29. Zhang, M., Chen, Y.: Link prediction based on graph neural networks. In: Advances in Neural Information Processing Systems, pp. 5171–5181 (2018)

The Diffusion of Vaccine Hesitation: Media Visibility Versus Scientific Authority

Zhai Yujia[1,2(✉)], Yao Yonghui[1], and Liang Yixiao[1]

[1] Management School, Tianjin Normal University, No. 393 Binshui West Road, Xiqing District, Tianjin 300387, China
zhaiyujiachn@gmail.com
[2] School of Information Management, Wuhan University, No. 16 Luojia Mountain, Wuchang District, Wuhan 430072, Hubei, China

Abstract. [**Purpose/Significance**] This study quantifies media visibility and scientific authority of vaccine scientists and anti-vaxxers. We analyze differences and associations through media co-occurrence and scientific inter-citation networks to comprehend vaccine hesitancy causes. [**Methods/Process**] We collected 100,000 research documents and 60,000 English-language media article metadata from 213 anti-vaxxers and 200 vaccine scientists. Differences in their media visibility were analyzed individually and as groups. We explored passive and active media presentation of anti-vaxxers and vaccine scientists. Co-occurrence and citation associations were examined through separate networks. Media articles were analyzed for frequency of appearance and pronoun use. [**Results/Conclusions**] Anti-vaxxers' media visibility is 52% higher, but top 50 vaccine scientists surpass anti-vaxxers in visibility. Media focus on anti-vaxxer topics drives attention. Despite limited scientific authority, anti-vaxxers gain traction through disinformation. Vaccine scientists gain visibility based on their scientific authority. Anti-vaxxers' close interconnections induce team effects, aiding opposition spread. Controversial nature makes anti-vaxxers more frequent in coverage. Pronoun differences highlight contrasting perspectives. These findings aid understanding of vaccine reporting and information dissemination for tackling vaccine hesitancy.

Keywords: Vaccine hesitancy · Scientific authority · Media visibility · Anti-Vaxxers · Vaccine Scientists · media co-occurrence networks · scientific inter-citation networks

1 Introduction

In early 2020, a global outbreak of the novel coronavirus (COVID-19) severely impacted the world's economy, healthcare, and daily lives. In this crisis, herd immunity emerged as a vital tool to manage the epidemic's spread and mutations [1]. Herd immunity safeguards vulnerable individuals indirectly by immunizing a significant portion of the population [2]. Vaccination stands as a critical defense against COVID-19 [3]. Nevertheless, the World Health Organization confirms that current global vaccination rates remain inadequate for achieving herd immunity.

F. Wu et al. (Eds.): SMP 2023, CCIS 1945, pp. 30–47, 2024.
https://doi.org/10.1007/978-981-99-7596-9_3

The Strategic Advisory Group of Experts on Immunization (SAGE) defines "vaccine hesitancy" as the choice to delay or refuse vaccination even when it's available; this varies based on factors like time, place, and the specific vaccine [4]. In 2019, WHO included vaccine hesitancy among the top 10 global health threats, alongside air pollution, antimicrobial resistance, Ebola, and others [5].· However, recent studies indicate that global vaccine hesitancy now imperils herd immunity [6], demanding swift attention within the context of the COVID-19 pandemic. Therefore, it's crucial to investigate the specific reasons behind COVID-19 vaccine hesitancy to develop effective strategies for addressing this issue.

Existing studies have pinpointed several factors impacting vaccine hesitancy amid the new crown pneumonia pandemic. These include the perceived severity and risks of COVID-19[7], perceptions of vaccine safety and effectiveness [8], worries regarding potential side effects of the New Coronavirus vaccine [9], and trust in local governments [10]. Yet, few studies have delved into the sway of online media and media articles on vaccine hesitancy.

In our rapidly evolving digital landscape, online media serves as a pivotal tool to transcend temporal and spatial limitations, providing swift access to pertinent information [11]. Particularly amid the hasty advancement and promotion of the COVID-19 vaccine, which remains relatively enigmatic, individuals are actively resorting to online media for comprehensive information about the vaccine. Consequently, information disseminated through online media significantly contributes to individuals' understanding of the COVID-19 vaccine.

Wilson highlighted that online misinformation stands as a prominent catalyst for vaccine hesitancy, underscoring the role digital platforms play in shaping public perspectives [12]. Additionally, media depictions of vaccines wield substantial influence over public sentiments. Germani's analysis of media portrayal revealed certain channels amplifying anti-vaccine voices, intensifying vaccine hesitancy [13].

Within the context of media prominence, Bucchi investigated how scientific experts' visibility in media affects public vaccine acceptance. Their findings suggest that heightened media attention and trust in scientists correlate with increased public vaccine acceptance [14]. This aligns with our study, which discovered that despite scientific credentials, vaccine scientists garner less media visibility compared to anti-vaxxers.

Moreover, the role of social media cannot be dismissed. Puri's research delved into how social media platforms exacerbate vaccine hesitancy, acknowledging their dual nature as both information conduits and misinformation breeding grounds [15]. Bajwa explored mainstream media's role in shaping public vaccine opinions, noting a focus on controversies and negativity, potentially distorting public perceptions [16].

This study, a comparative analysis of media prominence and scientific authority of anti-vaxxers and vaccine scientists at individual and group levels, demonstrates through quantitative analysis the media visibility bias towards anti-vaccinators. This suggests skewed media coverage for both groups, irrespective of scientific authority. Furthermore, the study assessed individuals' scientific authority by mapping associations in media co-occurrence and scientific mutual citation networks. Findings unveiled significant media visibility imbalance favoring anti-vaxxers. Consequently, this distortion undermines vaccine scientists' credibility and authority, hindering objective scientific discourse. Analyzing the results, the study reveals the imbalanced portrayal of vaccine-related topics in the media, shedding light on factors fueling widespread vaccine hesitancy. The study offers valuable insights for addressing this issue.

2 Data and Methods

2.1 Datasets

2.1.1 List of Anti-Vaxxer/Data

We conducted a thorough search on Wikipedia and Google News, focusing on articles containing the keyword "Anti-Vaccine". We meticulously sifted through each news article to identify and extract information about individuals and organizations associated with the anti-vaccine stance. Notably, our scope encompassed figures who are notably mentioned in the public domain, as evident by their presence in Wikipedia entries and frequent media exposure. This emphasis ensures a curated collection of 200 anti-vaxxers, spanning diverse professions such as academics, scientists, politicians, businessmen, and public figures, hailing from various Western countries in Europe and the United States.

Each entry in our compilation is meticulously documented, encompassing crucial details. For individuals, we record specific statements made, along with their professional identity, which could range from scientist to politician to celebrity. Similarly, for anti-vaccine organizations, we catalog data regarding their founding members, establishment dates, concise descriptions, and web addresses.

It's worth emphasizing that this assemblage is exclusive to individuals and entities that occupy a distinct public presence, substantiated by the existence of Wikipedia entries. This approach ensures that our collection reflects figures and organizations frequently in the public spotlight.

2.1.2 List of Vaccine Scientists/Data

The 200,000 references to Vaccine are collected from the WOS database and the 200 most frequently cited scientists are selected as a control group. In order to avoid information errors due to renaming, the relevant articles of individual scientists are also recorded and the WOS personal homepage URL is saved during the search.

2.1.3 Media Data

After obtaining the list of opposing sides, we consolidate the data collected into a database of print, online and blog articles on the topic of "Vaccine" from the Media Cloud (MC). MC is an open-source platform for studying the media ecosystem, hosted by the MIT Center for Civic Media and the Berkman Klein Center for Internet and Society at Harvard University, allowing researchers to track how stories and ideas spread through the media by tracking the millions of stories published online. Spread through the media, we aim to aggregate, analyses, deliver and visualize information to answer complex quantitative and qualitative questions about online media content.

Two types of queries are executed for this study:

1. Using the MC search query "Vaccine or Anti-Vaccine," we amassed 667,335 media articles.
2. Using the MC search query "member name and ** and Vaccine" (e.g. "Smith and Michael and Vaccine"), each individual from the study list was searched individually, resulting in 233 separate documents of varying lengths.

Each article sourced from the MC database possesses a unique identifier (IDA). Duplicates with the same IDA were eliminated, yielding a dataset exceeding 30,000 articles. It's notable that a subset of articles couldn't be accessed in full due to copyright restrictions. However, over 24,000 articles were obtained in full text.

The debate on vaccine hesitancy started in Europe and the US. Since the onset of COVID-19, the spread of vaccine information has faced increased challenges in these regions, with many Western countries having significant vaccine-hesitant populations. Given that 98% of the articles in the MC database are in English, utilizing English-language media data allows this study to discern differences in media visibility and scientific authority between positive and negative vaccine statements. This can help address challenges faced by positive statements and offer insights for effective vaccine communication in China.

2.1.4   Mainstream Media Sources

This study examines differences in media visibility and scientific authority between anti-vaxxers and vaccine scientists across various media platforms. We selected 10 mainstream media sources based on the volume of vaccine-related articles they published, as detailed in the table below (Table 1).

Table 1. Selection of ten mainstream media sources

No.	Media Name	Description
1	Washington Post	The second largest newspaper in the United States
2	New York Times	The Times is a daily newspaper published in New York, USA
3	USA Today	The only national color English language folio daily newspaper in the United States
4	Wall Street Journal	The largest paid circulation financial newspaper in the US
5	LA Times	Has the fifth largest circulation in the US and is the largest US newspaper not based on the East Coast
6	Seattle Times	The largest daily newspaper in Washington State, USA

(*continued*)

Table 1. (*continued*)

No.	Media Name	Description
7	Arkansas Democrat Gazette	Newspaper of record for Arkansas, USA
8	SF Chronicle	One of the largest circulated newspapers in the United States
9	New York Post	A conservative daily newspaper published in New York City
10	San Jose Mercury News	A morning newspaper published in the San Francisco Bay Area, San Jose, California, USA

2.2 Methods

This paper employs content analysis on media articles to examine the connection between media and vaccine hesitancy, specifically discerning factors that distinguish anti-vaxxers from vaccine scientists and also what binds them. For this endeavor, two extensive datasets were curated: 200,000 vaccine research articles from WOS and 24,000 English-language vaccine media pieces from the Media Cloud (MC) project. The study includes a comparative analysis of media visibility and scientific authority between 200 active anti-vaxxers and 200 vaccine scientists engaged in vaccine research.

2.2.1 Comparison of Media Visibility and Scientific Authority

We evaluate media visibility between the two groups by collecting the following data: (1) The 100 most prominent anti-vaxxers advocate in the media, ranked by the number of MC articles (Mi). (2) The leading media sources for all anti-vaccine articles, denoted by the total number of articles from these sources (Ms). (3) The 100 top-visible vaccine scientists in the media (Mi). (4) The most prolific media sources for articles related to vaccine scientists (Ms).

To assess the scientific authority of both anti-vaxxers and vaccine scientists, we analyzed their publication and citation counts. By matching individual names with the co-authors in the WOS dataset, we addressed potential author name disambiguation [17]. Notably, only 56 of the 200 anti-vaxxers appeared in the WOS dataset. For balanced analysis, we compared the publications of 50 selected anti-vaxxers with the most cited articles of 50 vaccine scientists, forming the 'Anti-Vaccine' and 'Vaccine' datasets. Literature co-authored by multiple dataset members was counted singularly to avoid double-counting.

Delving deeper into the media visibility differences between anti-vaxxers and vaccine scientists, we ranked individuals by their media article counts (Mi) using data from 10 prominent media sources, and created scatter plots comparing individual publications (Pi) to media mentions.

To discern how anti-vaxxers and vaccine scientists achieve media visibility, we scrutinized the full-text content of four key media outlets: New York Post (NYP), San Jose Mercury News, USA Today, and New York Times (NYT), analyzing 1,678 articles. Each article was evaluated to identify individuals and deduce their context-based roles. These individuals were subsequently categorized into five distinct groups. For comparative clarity, we streamlined these into two primary categories: passive mention (Mention) and active presentation (Contribution).

2.2.2 Media Co-occurrence Networks

Due to the extensive volume of media articles, it's infeasible to conduct content analysis on the entire dataset. Instead, this study employs network analysis to discern prevalent patterns of visibility relationships both within and among media outlets.

We started by merging media article data from both vaccine scientists and anti-vaxxers. For every person i, M_i represents their total media article count, while M_{ij} ($\leq M_i$) signifies articles that feature both individuals i and j. Calculating the M_{ij} matrix element for each individual, we fashioned a co-visibility matrix, M. It's crucial to mention that those with a $M_i = 0$ aren't part of matrix M. Conversely, those with a $M_i > 0$ are incorporated into M, even if they aren't co-mentioned with others in any articles.

2.2.3 Scientific Inter-citation Networks

This study also analyzes the organizational patterns observed in the WOS citation network. Citation networks are constructed from the reference lists of academic literature and represent a complex system that demonstrates the interactions between researchers, scholarly outputs, collective knowledge, and emerging cultures. Citation networks can provide a way to examine the evolution of scientific work [18]. Scientific authority is derived from repeated interactions between individuals in a community of active scientists, and can therefore be inferred from the total number of citations at different levels [19]. In the current context, the different citations represent quantifiable interactions between individuals, which may range from attribution to critique to outright dismissal. If the latter type of negative citation occurs relatively frequently, it reflects the adversarial nature of the debate around controversial scientific issues [20, 21].

2.2.4 Textual Analysis

Through a textual analysis of media articles, we extracted pronouns representing individuals and determined their frequency and relative occurrence. This approach sheds light on the differing media portrayals of anti-vaxxers and vaccine scientists, offering insights into the sway of anti-vaccine sentiments in public discourse. Complementing this, we employed natural language processing tools, including the LIWC thesaurus and named entity recognition, to evaluate the articles. Key textual segments were extracted, and their frequency and proportionate presence in the articles were computed.

For our analysis, we utilized the 2015 edition of the LIWC lexicon, alongside the Spacy text analysis method implemented in Python. Our focus revolved around specific pronouns such as "you", "I", "he/she", and words that prompt direct questions to the audience [22]. By aligning these keywords with the full text of media articles, we discerned their frequency and relative prevalence.

Furthermore, we incorporated the pre-trained and verified named entity recognition feature of the Python spaCy NLP library, specifically employing the "en_core_news_lg" model tailored for English news articles. This enabled the identification of words denoting individuals. From 24,000 media pieces, we quantified pronouns linked to individual names, capturing those linked to both proponents and adversaries of vaccination. This informed us about the proportionate personalized narratives present in these articles.

Personalization was a significant aspect of our study. We gauged the level of individual-centric narratives in media pieces by pinpointing descriptive words related to people. Such evaluations bestow a thorough and unbiased insight into the representation and depiction of anti-vaxxers and vaccine scientists in the media. It aids in grasping the proliferation and sway of anti-vaccine sentiments.

Moreover, the presentations of the way individuals are engaged in media articles fall into five categories: (1) Reference to a person by name only; (2) A quotation from a person's relevant scientific statements that elaborates on certain issues; (3) A quotation from a person's neutral statements without any additional interpretation; (4) Similar to the second category, quoting a person's statements but addresses confrontational statements from the other side; (5) A media article written by an individual.

3 Results

3.1 Building Datasets Around Individuals

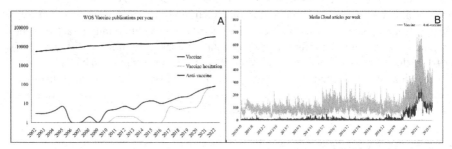

Fig. 1. Growth in research output and media production related to vaccines. A: Vaccine WOS publications per year. B: Media Cloud articles per week

Over the past two decades, research and reporting on vaccines has continued to grow, particularly in the last three years, due to the impact of COVID-19. This growth has attracted experts from a wide range of fields. As of May 2022, according to the MC data report (Fig. 1B), the term "vaccine" appeared in media articles an average of 36,032 times per week, and the 'anti-vaccine' approximately 100 times more often than the it. The term 'anti-vaccine' is a broad term referring to anti-vaxxers and members of anti-vaccine groups. Meanwhile, this study delves further into their social media presence on platforms such as Twitter and Facebook. It finds that they repeatedly and publicly state their opposition and resistance to vaccines in various social media posts, as well as spreading anti-vaccine-related rhetoric.

3.2   Comparison of Media Visibility

Figure 2 shows the 100 most prominent anti-vaxxers and vaccine scientists in the media, and the 45 most prominent media sources. The difference in media visibility between the two groups can be seen more clearly in the graph. Figures 2(A) and 2(C) represent the ranking of the top 100 anti-vaxxers and vaccine scientists in terms of the number of media articles. This provides a visual representation of the change in media visibility of vaccine scientists and anti-vaxxers and enables the identification of the most active vaccine scientists and anti-vaxxers on the topic. Figure 2(B) and Fig. 2(D) represent the ranking of the number of articles published by the media related to vaccine scientists and anti-vaxxers, respectively. These figures can help identify the most prominent media sources on vaccine-related topics. To protect privacy, individuals on the list are anonymized in this study.

Specifically, the average number (median) of media articles for the top 100 anti-vaxxers is 221 (143.5); similarly, for vaccine scientists, the average number (median) is 32 (21). It is clear from this study that anti-vaxxers have much greater media visibility than vaccine scientists, reflecting the fact that media pay much more attention to anti-vaxxers than vaccine scientists when publishing vaccine-related articles. This

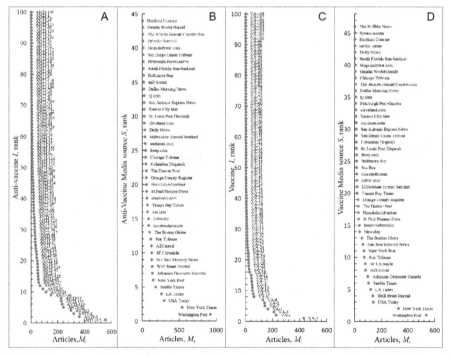

Fig. 2. Comparison between anti-vaxxers and vaccine scientists in the media. A: The 100 most prominent anti-vaxxers in the media.and ranked based on the number of MC articles (M_i). B: The most prolific media sources among all anti vaccine articles, represents the total number of articles from media sources (M_S). C: The 100 most prominent vaccine scientist in the media(M_i). D: The most prolific media sources among all vaccine scientist articles(M_S).

can contribute to the increased attention that anti-vaxxers receive, which may increase people's vaccine hesitancy.

3.3 Comparison of Scientific Authority

A comparison of the literature and citation volume of the 50 selected anti-vaxxers and vaccine scientists is shown in Fig. 3. Figure 3A illustrates the difference in scientific productivity between the two groups, with Anti-Vaccine publishing 6,554 publications (an average of approximately 131 citations per person), while Vaccine published approximately twice as many with 14,391 publications. Figure 3B shows that Vaccine has 23 times more citations than Anti-Vaccine, indicating a much greater difference in citation impact between the two groups. These results suggest that there is a clear gap between anti-vaxxers and vaccine scientists in terms of scientific authority. It can be assumed that most anti-vaxxers do not have authority in the field of vaccine research and that their expertise may not support the science and rigor of their statements.

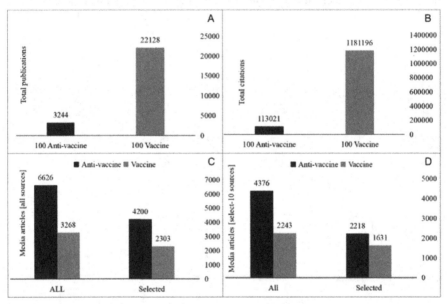

Fig. 3. Differences in scientific authority and media visibility-group level A: Total number of publications. B: The total number of citations for publication. C: Number of media articles from all media sources. D: Select the number of media from 10 mainstream media sources.

When we again compare the media visibility of anti-vaxxers and vaccine scientists based on a selection of 50 anti-vaxxers and vaccine scientists, the media visibility results are similar (2,457 media data for vaccine scientists and 2,434 media data for anti-vaxxers). Again, when counting within the Select-10 range, vaccine scientists are found to have a 17% media visibility advantage (718 media data for anti-vaccine activists, 809 media data for vaccine scientists). That is, the media visibility of vaccine scientists gains a degree of advantage in the mainstream media only when the scientific authority is much higher than that of anti-vaxxers. It can also be concluded that the media visibility of vaccine scientists is more concentrated among the more scientifically authoritative segment of scientists than among anti-vaxxers.

To summarize, the overall media visibility advantage of anti-vaxxers may be due to the large amount of attention that the topic of "anti-vaccine" has received in the context of the current epidemic. When deciding what to report, the media often tends to favor the side with higher scientific authority or the side that generates the most attention. However, most vaccine scientists are not well-known to the general public, and only a small number of authoritative experts in the field of vaccines are recognized by the public and attract attention. Conversely, despite the weak scientific authority of many anti-vaxxers, they can still generate significant interest through false facts and exaggerated propaganda on various social media platforms or in other new media.

3.4   Scientific Authority and Media Visibility at the Individual Level

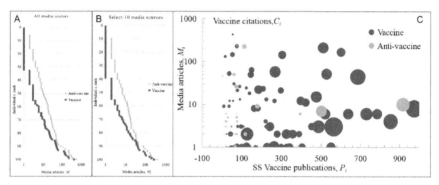

Fig. 4. Differences between scientific authority and media visibility in individual level. A: Rank individuals based on the number of media articles M_i. B: Calculated from 10 mainstream media sources M_i. C: Individual scatter plots of publications P_i and media visibility M_i, with the size proportional to the total citation volume C_i

Figure 4A shows the ranking of the number of media articles (Mi) published by individuals, allowing for comparisons of article counts between individuals with the same ranking within the group. The data reveals that anti-vaxxers consistently have a higher number of media articles compared to vaccine scientists. This difference persists even when comparing visibility within media sources on the Select-10.

Figure 4B shows the total number of media articles (M$_i$), the total WOS literature (P$_i$) and the total number of citations received (C$_i$). Although anti-vaxxers dominate in terms of overall media visibility, only a minority of anti-vaxxers can match the scientific achievements of vaccine scientists. Furthermore, anti-vaxxers are more likely to have greater Mi values than vaccine scientists for the same range of literature numbers(Pi). Thus, despite the more significant scientific dominance of vaccine scientists, the difference in media visibility between the two groups is significant.

3.5 Ways to Engage with Media Visibility

Figure 5 (left) shows the frequency distribution of the five types of presentation, whereas Fig. 5 (right) shows the distribution of the two main types of presentation. As seen in Fig. 5 (b), the four media sources report more articles about anti-vaxxers. Vaccine scientists tend to be more frequently presented in an active manner, whereas anti-vaxxers

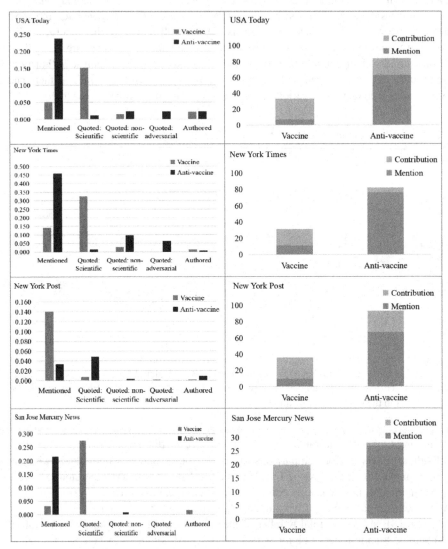

Fig. 5. Presentation of anti-vaxxers and vaccine scientists in the media. Mention: Only mention someone's name; Scientific: Citing someone's relevant scientific remarks, which is an explanation of certain issues; Non-scientific: Quoting someone's neutral statement without any additional interpretation; Adversarial: refers to a statement made by someone, but it is a confrontational statement made against the other party; Authored: Media articles written by individuals.

are mentioned more often in media articles. As can be seen in Fig. 6(a), in the San Jose Mercury News, USA Today, and the New York Times, vaccine scientists mostly present their scientific content actively in media articles (i.e., the second type of presentation), whereas anti-vaxxers are mentioned passively in media articles (i.e., the first type of presentation). In the New York Post, however, vaccine scientists and anti-vaxxers are presented in the opposite way to the three media sources mentioned above.

As mentioned earlier, the first type of presentation does not necessarily imply that the individual mentioned has contributed positively to the content of the article, nor is the content considered scientific or authoritative. The second type of presentation involves more citations of content from vaccine scientists and is therefore more common among vaccine experts, and rarely associated with anti-vaxxers. Among the five types of presentation, the least common are non-scientific content quotes and adversarial quotes, with the majority of these types being more prominent among anti-vaxxers. This suggests that the media often reports anti-vaxxers' statements that lack scientific validity and authority. The media may present these statements to provide balance or to offer alternative views on statements that refute anti-vaxxers.

3.6   Media Co-occurrence Networks

Fig. 6. Media Co-occurrence Network

The co-occurrence matrix has two basic characteristics: First, anti-vaxxers are 60% more prominent in the co-visibility matrix than vaccine scientists. The advantage in visibility for anti-vaxxers increases to 68% when co-occurrence in Select-10 media sources is considered. Second, the strongest co-occurrence instances (the largest Mij) occur within groups rather than between groups.

Specifically, this study applies the modularity algorithm to identify the communities within the network, and the results indicate that individuals within a community have more ties than those outside [23]. As depicted in Fig. 6, which visualizes each community on a separate axis, there exist three distinct community structures. Moreover, within each axis (or community), individuals (or nodes) are ranked based on their network centrality. The Page Rank metric is used for this ranking, positioning the most influential individuals of each community at the top.

By examining the composition of each community, it is possible to categorize them into two types: two of the communities are a mixture of the two groups, while the third community consists mainly of anti-vaxxers - a classic example of the echo chamber effect, whereby anti-vaxxers are more likely to be in an insular community. Vaccine scientists, on the other hand, are much less aggregated, and the echo chamber effect tends to make the rhetoric of anti-vaxxers more believable to the masses.

3.7 Scientific Inter-citation Networks

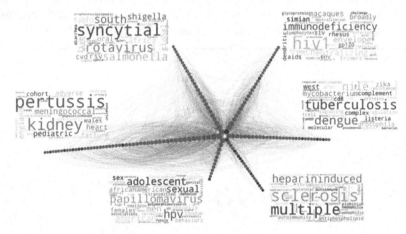

Fig. 7. Scientific Authority Network

Therefore, this study assesses the flow of authority between anti-vaxxers and vaccine scientists at the group and individual levels through the citation network of "Vaccine". Directing citation flow between individuals, i.e. the citation flow that occurs when a paper P_a authored by individual A cites a paper P_b authored by individual B. Specifically, this study computes citation links $A{\sim}P_a \rightarrow P_b{\sim}B$ connecting any pair of authors, where $A{\sim}P_a$ indicates that individual A is the author of literature P_a and $P_a \rightarrow P_b$ while indicating that literature P_a cites P_b.

Figure 7 displays the inter-citation network between individuals, also known as the scientific inter-citation network, where only 22 anti-vaccine individuals are connected. The results show that vaccine proponents are connected to each other 3,437 times (87.3%), vaccine opponents are connected to each other 481 times (12.2%), and anti-vaxxers are only connected to each other 16 times (0.4%). Additionally, anti-vaxxers cite vaccine scientists with approximately the same frequency as they are cited by them.

The network consists of six communities, each represented as an axis, with individuals ranked according to centrality, and the most authoritative individuals located at the top. Additionally, this study uses the TF-IDF algorithm to extract keywords from the abstracts of papers contained in each community and to annotate the community content. The scientific authority network indicates that the majority of anti-vaxxers are located in the periphery of the network. Interestingly, peripheral anti-vaxxers within each community tend to direct the majority of their citations to the most prominent vaccine scientists. This citation pattern may suggest adversarial interactions, including negative citations aimed at discrediting the findings of prominent vaccine scientists.

3.8   Media Full-Text Text Analysis

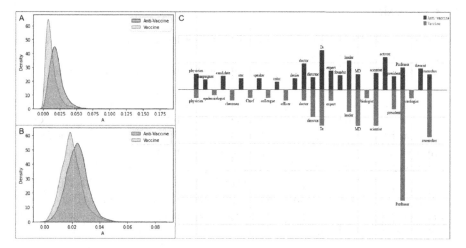

Fig. 8. Full text analysis of media articles

The full text analysis of media articles, as illustrated in Fig. 8, elucidates the portrayal of anti-vaxxers and vaccine scientists. Media articles often depict anti-vaxxers with negative descriptors like "anti-science" and "suspicious." Such descriptions might heighten readers' interest but can also amplify controversy. Due to their polarizing views and actions, anti-vaxxers often secure more media limelight than experts and scholars. The media's inclination towards highlighting contentious figures can be attributed to its aim of boosting readership and consequently, ad revenues. Vaccine scientists, majorly proponents of vaccination and bearers of mainstream opinions, may experience comparatively diminished media exposure. Nonetheless, when discussing experts, media does employ sensational terms to captivate readers. It's possible that the media's extensive coverage of anti-vaxxers stems from their controversial nature rather than a deliberate oversight of expert scholars.

Delving into the analysis of character pronouns from media articles, pronounced disparities were observed in the descriptors for anti-vaxxers versus vaccine scientists, perhaps mirroring the media's framing and intended audience perception.

Media stories concerning vaccine scientists often use titles such as "Dr" and "MD," underscoring their professional stature in medicine. Amidst a pandemic, statements from medical experts bear significant weight, leading media outlets to accentuate their qualifications and credibility. Pronouns like "professor," "scientist," and "researcher" are also prevalent, indicating the media's recognition of the critical role played by research and academia during health crises.

Conversely, when addressing anti-vaxxers, the media gravitates towards more disparaging or skeptical titles such as "activist," "denier," and "theorist." This could suggest that the media perceives anti-vaxxer opinions as less evidence-based and more speculative, and these titles are strategically used to capture audience attention. Titles denoting authority like "leader" and "president" also frequently surface, hinting at the media's focus on the hierarchical and influential dynamics within anti-vax circles..

4 Conclusion

The research analyzes the disparities in media coverage between anti-vaxxers and vaccine scientists, revealing how media may inadvertently or deliberately bolster anti-vaccine sentiments, potentially diluting the more authoritative voices of vaccine experts.

The findings of this study indicate that anti-vaxxers enjoy 52% more media exposure than vaccine scientists. Even in mainstream outlets that heavily feature vaccine-related content, anti-vaxxers outshine the experts in visibility. Yet, when spotlighting the top 50 figures from both groups, vaccine experts regain visibility, leading by 17%. The experts also dwarf anti-vaxxers in terms of scholarly contributions, with 219% more publications and a staggering 2,396% advantage in citations.

Analyzing 3,000 articles from prominent outlets like USA Today, NYP, NYT, and San Jose Mercury News, the study showed that anti-vaxxers often get mentioned or are cited non-scientifically. Vaccine scientists, conversely, typically contribute via scientific citations or as article authors. Journalists' propensity to quote anti-vaxxers might be rooted in the pursuit of journalistic neutrality or to highlight the fallacies in anti-vax arguments. Yet, this style of coverage inadvertently shadows significant vaccine developments and might suggest to readers that anti-vaccine views have substantial merit.

When exploring citation networks, vaccine scientists out-cite each other 170 times more frequently than anti-vaxxers do amongst themselves. However, when it comes to reverse citations, the top 50 figures from both groups are cited almost equally. Analyzing media co-occurrence patterns, 54% happened between the two groups, with anti-vaxxers dominating 34% of co-occurrences, leaving only 8% for the vaccine experts. Structural differences, such as the echo chamber phenomenon, further differentiate the two groups.

In summary, this study employs large-scale data collection and analysis to examine individual and collective characteristics associated with intricate societal matters. It offers valuable insights into the polarization surrounding contentious socio-political matters, the impact of the media on the public, and the diffusion of inaccurate information. These findings can serve as a valuable resource for enhancing the effectiveness of positive vaccine communication.

5 Discussion

Vaccine hesitancy has been somewhat mitigated by the novel coronavirus (COVID-19) pneumonia outbreak, but distrust of vaccines remains and persists, and predictions of vaccine hesitancy are not encouraging. Online media, as the primary information source for many, can inadvertently promote this hesitancy, jeopardizing public health. The rise of anti-vaxxer narratives online paints vaccines in negative lights, focusing on alleged "corruption" and safety risks. Misinterpretations of official findings and expert views further confuse the public, potentially leading to vaccination delays [24]. The public tends to view information through personal biases [25]. Rather than directly engaging with experts, they often receive diluted opinions through media [26]. As media tries to balance perspectives, including anti-vaxxer views [27], it poses challenges in conveying accurate vaccine information.

The spread of credible vaccine information is vital in combating global epidemics. However, consistently delivering authoritative details is tough. One challenge is experts often simplify specialized knowledge for the general public to avoid misunderstandings [28]. Moreover, the public's demand for expertise is dwindling [29]. The rise of online media complicates this, making it easy to access but hard to discern credible sources. Hence, we must improve our ways of communicating vaccine expertise to the public.

This study presents some limitations in data analysis. We did not differentiate various vaccine hesitancy factors, such as the recent distinction between basic skepticism (on vaccine safety) and impact skepticism (on vaccine-related event consequences). The shifting dynamics of these skepticisms suggest evolving anti-vaccine narratives. Distinguishing between them might offer a more granular view of individual media visibility. This distinction and its portrayal in various media is a potential area for further research.

The study encompasses anti-vaxxers from varied backgrounds, including scientists, entrepreneurs, politicians, and liberal-leaning individuals. This diverse group can disseminate information ranging from misleading to profit-driven narratives. To mitigate this, we used Wikipedia's definition of anti-vaxxers to ensure data accuracy. We also introduced a control group of vaccine scientists, matching its size with the anti-vaxxer group.

We didn't differentiate between pro-vaccine individuals advocating for science and those countered as misinformation spreaders. Thus, our media visibility metric might be skewed by negative mentions, akin to the effect of positive vs. negative citations.

To further augment the study in the future, two avenues can be explored. First, incorporating more data from mainstream media's social platforms and articles will offer a broader viewpoint on the topic. Second, leveraging natural language processing can help refine the presentation of vaccine scientists' findings and counter anti-vax narratives.

Acknowledgement. This research was funded by the Major Program of National Social Science Foundation of China, grant number 20&ZD141.

References

1. Fontanet, A., Cauchemez, S.: COVID-19 herd immunity: Where are we? Nat. Rev. Immunol. **20**(10), 583–584 (2020). https://doi.org/10.1038/s41577-020-00451-5
2. Randolph, H.E., Barreiro, L.B.: Herd immunity: understanding COVID-19. Immunity **52**(5), 737–741 (2020). https://doi.org/10.1016/j.immuni.2020.04.012

3. Yuan, H., Long, Q., Huang, G., Huang, L., Luo, S.: Different roles of interpersonal trust and institutional trust in COVID-19 pandemic control. Soc. Sci. Med. **293**, 114677 (2022). https://doi.org/10.1016/j.socscimed.2021.114677
4. MacDonald, N.E.: Vaccine hesitancy: Definition, scope and determinants. Vaccine **33**(34), 4161–4164 (2015). https://doi.org/10.1016/j.vaccine.2015.04.036
5. Pugazhenthan, T., Sajitha, V.: WHO Ten threats to global health in 2019: Antimicrobial resistance. Cukurova Med. J. **44**(3), 1150–1151 (2019). https://doi.org/10.17826/cumj.514157
6. Calnan, M., Douglass, T.: Hopes, hesitancy and the risky business of vaccine development. Health Risk Soc. **22**(5–6), 291–304 (2020). https://doi.org/10.1080/13698575.2020.1846687
7. Carcelen, A.C., et al.: COVID-19 vaccine hesitancy in Zambia: A glimpse at the possible challenges ahead for COVID-19 vaccination rollout in sub-Saharan Africa. Hum. Vaccin. Immunother. **18**(1), 1–6 (2022). https://doi.org/10.1080/21645515.2021.1948784
8. Yeşiltepe, A., Aslan, S., Bulbuloglu, S.: Investigation of perceived fear of COVID-19 and vaccine hesitancy in nursing students. Hum. Vaccin. Immunother. **17**(12), 5030–5037 (2021). https://doi.org/10.1080/21645515.2021.2000817
9. Hossain, E., et al.: COVID-19 vaccine-taking hesitancy among Bangladeshi people: Knowledge, perceptions and attitude perspective. Hum. Vaccin. Immunother. **17**(11), 4028–4037 (2021). https://doi.org/10.1080/21645515.2021.1968215
10. Arvanitis, M., et al.: Factors associated with COVID-19 vaccine trust and hesitancy among adults with chronic conditions. Prev. Med. Rep. **24**, 101484 (2021). https://doi.org/10.1016/j.pmedr.2021.101484
11. Zhou, C., Xiu, H., Wang, Y., Yu, X.: Characterizing the dissemination of misinformation on social media in health emergencies: An empirical study based on COVID-19. Inf. Process. Manage. **58**(4), 102554 (2021). https://doi.org/10.1016/j.ipm.2021.102554
12. Wilson, Steven Lloyd, Wiysonge, Charles: Social media and vaccine hesitancy. BMJ Global Health **5**(10), e004206 (2020)
13. Germani, F., Biller-Andorno, N.: The anti-vaccination infodemic on social media: A behavioral analysis. PLoS ONE **16**(3), e0247642 (2021)
14. Bucchi, M., Fattorini, E., Saracino, B.: Public perception of COVID-19 vaccination in Italy: the role of trust and experts' communication. Int. J. Public Health **67**, 1604222 (2022)
15. Puri, N., Coomes, E.A., Haghbayan, H., Gunaratne, K.: Social media and vaccine hesitancy: new updates for the era of COVID-19 and globalized infectious diseases. Hum. Vaccin. Immunother. **16**(11), 2586–2593 (2020)
16. Bajwa, A.: Information disorder, the Triumvirate, and COVID-19: How media outlets, foreign state intrusion, and the far-right diaspora drive the COVID-19 anti-vaccination movement. J. Intel. Conflict Warfare **4**(2), 16–45 (2021)
17. Milojević, S.: Accuracy of simple, initials-based methods for author name disambiguation. J. Informet. **7**(4), 767–773 (2013). https://doi.org/10.1016/j.joi.2013.06.006
18. Fortunato, S., et al.: Science of science. Science **359**(6379), eaao0185 (2018). https://doi.org/10.1126/science.aao0185
19. Petersen, A.M., et al.: Reputation and impact in academic careers. Proc. Natl. Acad. Sci. **111**(43), 15316–15321 (2014). https://doi.org/10.1073/pnas.1323111111
20. Catalini, C., Lacetera, N., Oettl, A.: The incidence and role of negative citations in science. Proc. Natl. Acad. Sci. **112**(45), 13823–13826 (2015). https://doi.org/10.1073/pnas.1502280112
21. Edelmann, A., Moody, J., Light, R.: Disparate foundations of scientists' policy positions on contentious biomedical research. Proc. Natl. Acad. Sci. **114**(24), 6262–6267 (2017). https://doi.org/10.1073/pnas.1613580114

22. Haim, M., Karlsson, M., Ferrer-Conill, R., Kammer, A., Elgesem, D., Sjøvaag, H.: You should read this study! It investigates Scandinavian social media logics ₫. Digit. Journal. **9**(4), 406–426.https://doi.org/10.1080/21670811.2021.1886861

23. Petersen, A.M., Majeti, D., Kwon, K., Ahmed, M.E., Pavlidis, I.: Cross-disciplinary evolution of the genomics revolution. Science Advances **4**(8), eaat4211 (2018). https://doi.org/10.1126/sciadv.aat4211

24. Miko, D., Costache, C., Colosi, H.A., Neculicioiu, V., Colosi, I.A.: Qualitative assessment of vaccine hesitancy in Romania. Medicina **55**(6), 282 (2019). https://doi.org/10.3390/medicina55060282

25. Kahan, D.M., et al.: The polarizing impact of science literacy and numeracy on perceived climate change risks. Nat. Clim. Chang. **2**(10), 732–735 (2012). https://doi.org/10.1038/nclimate1547

26. Wilson, C.C., Gutierrez, F.F.: Race, Multiculturalism, and the Media: From Mass to Class Communication, July 25 1995 (1995). https://www.semanticscholar.org/paper/Race%2C-Multiculturalism%2C-and-the-Media%3A-From-Mass-to-Wilson-Gutierrez/2a7743396639e12c6be5c7ba5d9459bd6db1efad

27. Dunwoody, S.: Weight-of-evidence reporting: What is it? Why use it? Nieman Rep. **59**(4), 89 (2005)

28. Bruine De Bruin, W., Bostrom, A.: Assessing what to address in science communication. Proc. Natl. Acad. Sci. **110**(supplement_3), 14062–14068 (2013). https://doi.org/10.1073/pnas.1212729110

29. Barclay, D.: Fake News, Propaganda, and Plain Old Lies: How to Find Trustworthy Information in the Digital Age, 25 June 2018 (2018). https://www.semanticscholar.org/paper/Fake-News%2C-Propaganda%2C-and-Plain-Old-Lies%3A-How-to-Barclay/8996d59e6ad9793b1853e092f2a3bbd191dc68bd

An Emotion Aware Dual-Context Model for Suicide Risk Assessment on Social Media

Zifang Liang, Dexi Liu$^{(\boxtimes)}$ ⓘ, Qizhi Wan ⓘ, Xiping Liu ⓘ, Guoqiong Liao ⓘ, and Changxuan Wan ⓘ

School of Information Management, Jiangxi University of Finance and Economics, Nanchang 330032, China
dexi.liu@163.com, wanchangxuan@263.net

Abstract. Suicide risk assessment on social media is an essential task for mental health surveillance. Although extensively studied, existing works share the following limitations, including (1) insufficient exploitation of Non-SuicideWatch posts, and (2) ineffective consideration of the fine-grained emotional information in both SuicideWatch and Non-Suicide-Watch posts. To tackle these issues, we propose an emotion aware dual-context model to predict suicide risk. Specifically, SuicideWatch posts that contain psychological crisis are leveraged to obtain the suicidal ideation context. Then, given that suicidal ideation is not instantaneous and Non-SuicideWatch posts can provide essential information, we encode the emotion-related features and emotional changes with variable time intervals, revealing users' mental states. Finally, the embeddings of SuicideWatch, Non-SuicideWatch, LIWC feature, and posting time are concatenated and poured into a fully-connected network for suicide risk level recognition. Extensive experiments are conducted to validate the effectiveness of our method. In results, our scheme outperforms the first place in CLPsych2019 task B by 4.9% on Macro-$F1$ and achieves a 10.7% increase on $F1$ of Severe Risk label than the first place in CLPsych2019 task A that only uses SuicideWatch posts.

Keywords: suicide risk · emotion aware · dual-context · social media

1 Introduction

Suicide is a serious global public health issue. In 2021, World Health Organization reported that about 700,000 people committed suicide worldwide every year [1], and suicide is the second factor causing death in the age of 14–35 in the US [2]. Identifying early warning signals for suicidal people can lead to timely treatment. Also, further user-level recognition of suicide risk is conducive to assigning the medical resource and providing targeted services [3]. With the development of social media, people are more willing to post comments about their emotions and mental states on it, providing extensive practical data and facilitating the suicide risk assessment by effective analysis and exploitation [4].

F. Wu et al. (Eds.): SMP 2023, CCIS 1945, pp. 48–62, 2024.
https://doi.org/10.1007/978-981-99-7596-9_4

At present, social media data has been widely used to handle the suicide risk assessment task in existing work. Mathur et al. [5] and Sinha et al. [6] implemented suicide ideation on the post-level. Sawhney et al. [7] explored the recognition of user's latest tweet. Regarding user-level, the related methods are limited. Guar et al. [8] identified users into five-levels of suicide risk. Ren et al. [9] assessed the suicide risk for blog authors by encoding blog emotions and emotion intensities. In these studies, the mixed posts that contain Suicide-Watch and Non-SuicideWatch posts were exploited, where the SuicideWatch (SW) refers to the subreddit of Reddit committing the SucideWatch posts and Non-SuicideWatch (NSW) represents the posts published in other subreddits. Subsequently, Mohammadi et al. [10] merely focused on the SW posts. Matero et al. [11] separately modeled SW and NSW posts. However, there are still some limitations in the user-level assessment task of suicide risk.

First, the NSW posts are insufficient exploitation, such as the explicit emotional information in it. Intuitively, it is difficult to determine the level of suicide risk according to the SW post *"I'm tired of trying and I can't keep going*. Nevertheless, the information (e.g., mental health concerns or life trivia) posted in NSW can provide some essences for the assessment task. For example, the emotional contents (e.g., *"hate"*, *"tired of living"* and *"anxious"*) in the NSW post *"I hate myself...I'm tired of living in this box with 5 people...I'm just an anxious scared mess..."* are negative, and they reveal the opinions and status of the user. Therefore, the probability of identifying suicidal ideation will be increased.

Second, the emotional changes of users over time have not been fully developed. Existing studies adopted emotion lexicon and statistical methods to measure user's emotional change, losing the linguistic information of words and failing to accurately reveal the emotion in posts. Also, user's posting intervals are not uniform, and varying time intervals have not been considered when encoding emotional changes. Given the time-sensitive of emotions, the importance of emotional information at different intervals varies.

To address these limitations, we propose an Emotion-aware Dual context framework for Suicide Risk assessment, named SRED. Specifically, given the excellent performance of BERT (Bidirectional Encoder Representations from Transformers) [12], we adopt it to obtain the semantic meaning of posts and fine-tune a Plutchik BERT [13] with Plutchik-2 dataset [14] to explicitly encode the semantics of emotional information. Then, to obtain the sequence information of SW posts, the Bi-LSTM is exploited. Since common LSTM does not better handle the information of time interval, a Time-Aware LSTM (T-LSTM) is developed to model the dynamic movement of emotions across NSW posts, demonstrating the mental health states of users. Finally, we concatenate the embeddings of SW and NSW posts, as well as the representations of the LIWC (Linguistic Inquiry and Word Count) and the posting time, and then pour them into a multi-layer perceptron for user-level suicide risk assessment.

To sum up, the main contributions of this paper are as follows:

- By exploiting the emotion cues in NSW posts, we develop an emotion-aware dual context framework for user-level suicide risk assessment, which can handle the representations of SW and NSW posts according to their different characteristics.

- To reveal the emotional changes and mental states of users over times, we design a T-LSTM model to integrate the time intervals in NSW posts. This paper is the first work using the time information in user-level suicide risk assessment.

- Extensive experiments are conducted, and the results confirm the effectiveness of our model in assessing user-level suicide risk task. Our model outperforms the first place in CLPsych2019 task B with a superiority of 4.9% on $F1$ and achieves 10.7% higher than the first place in CLPsych2019 task A on $F1$ of Severe Risk (SR) label.

2   Related Work

In the early years, feature engineering methods were mainly employed. The features such as online behavior [15], psycholinguistic lexicons [16,17], and emotion dictionary [18] were used for mental issue recognition on social media [19–22].

Recently, deep neural network models [22–26] were concerned and applied to the assessment of psychological crisis. Yates et al. [27] adopted the convolutional neural network to encode depression and self-harming tendencies. To capture long-term dependencies in sequences, Cao et al. [28] exploited the LSTM to detect users' potential suicide risk on microblogs.

Subsequently, user contexts (e.g., emotional changes and historical posts) were explored to understand the forming process of user's psychological crisis [5,9,29–31]. Sinha et al. [6] developed a stacked ensemble model to identify the suicidal ideation for tweets by exploiting the historical tweets of a user along with their social graph. Cao et al. [30] adopted emotion lexicon to measure the intensities of different emotions and used emotional changes to detect user's suicide risk. Sawhney et al. [32] realized the suicidal ideation detection for the latest tweet based on the historical tweet context.

Further, to refine the granularity of suicide risk and target users, some studies on user-level can be found in [8,33,34]. By modeling user posts that hybridize SW and NSW posts, Guar et al. [8] argued a contextual CNN (C-CNN) to identify suicide risk severity levels of users. Mohammadi et al. [10] only focused on the SW posts and performed well in user-level suicide risk assessment. Moreover, the post information of SW and NSW was separately encoded by BERT and fed into an attention-based RNN model for suicide risk assessment in [11].

In summary, existing work has three limitations: (1) the fine-grained emotions in both SuicideWatch (SW) and Non-SuicideWatch (NSW) posts are not fully exploited; (2) the methods separately modeling the two types of posts is straightforward, ignoring their unique characteristics; (3) users' emotional changes over time are not considered to demonstrate the mental states of users.

3   Methodology

Following Task B of CLPysch2019 Shared Task, this paper aims to assess user suicide risk level based on the input posts; that is, a user is classified into one of four levels: *No Risk*(NR), *Low Risk*(LR), *Moderate Risk*(MR), *Severe Risk*(SR).

Given the i-th user u_i, the SW posts are denote as $P_{\mathrm{SW}}^{u_i} = \{(p_{\mathrm{SW}_1}^{u_i}, t_{\mathrm{SW}_1}^{u_i}), \cdots, (p_{\mathrm{SW}_j}^{u_i}, t_{\mathrm{SW}_j}^{u_i}), \cdots, (p_{\mathrm{SW}_n}^{u_i}, t_{\mathrm{SW}_n}^{u_i})\}$, where n denotes the number of SW posts, $(p_{\mathrm{SW}_j}^{u_i}, t_{\mathrm{SW}_j}^{u_i})$ refers to the j-th SW post $p_{\mathrm{SW}_j}^{u_i}$ published at time $t_{\mathrm{SW}_j}^{u_i}$. Similarly, $P_{\mathrm{NSW}}^{u_i} = \{(p_{\mathrm{NSW}_1}^{u_i}, t_{\mathrm{NSW}_1}^{u_i}), \cdots, (p_{\mathrm{NSW}_j}^{u_i}, t_{\mathrm{NSW}_j}^{u_i}), \cdots, (p_{\mathrm{NSW}_m}^{u_i}, t_{\mathrm{NSW}_m}^{u_i})\}$ are NSW posts, where m denotes the number of NSW posts, $(p_{\mathrm{NSW}_j}^{u_i}, t_{\mathrm{NSW}_j}^{u_i})$ is the j-th NSW post $p_{\mathrm{NSW}_j}^{u_i}$ released at $t_{\mathrm{NSW}_j}^{u_i}$ time.

In the following, we describe our SRED framework as shown in Fig. 1. It includes five components: (1) the Post Embedding, for initializing the embeddings of input posts; (2) the Bi-LSTM Layer, handling the features in SW posts; (3) the T-LSTM, functional in obtaining the emotional and temporal information in NSW posts; (4) User features, extracting the word frequency in LIWC dictionary and posting behaviors of users; (5) Classification Layer, for predicting the level of suicide risk.

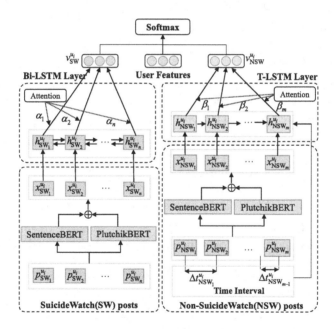

Fig. 1. SRED model architecture.

3.1   Post Embedding

Given the excellent performance of BERT pre-trained language model, we adopt the SentenceBERT to obtain the embeddings of the input posts. For user u_i, the embeddings of j-th SW and NSW post can be written as:

$$s_{\text{SW}_j}^{u_i} = \text{SentenceBERT}(p_{\text{SW}_j}^{u_i}) \tag{1}$$

$$s_{\text{NSW}_j}^{u_i} = \text{SentenceBERT}(p_{\text{NSW}_j}^{u_i}) \tag{2}$$

Due to the complex and diverse emotions expressed in posts, we fine-tune the BERT model based on the Plutchik-2 dataset constructed by Abdul-Mageed and Ungar [14], forming the PlutchikBERT model, which can demonstrate the fine-grained emotions in posts (e.g., eight emotions: joy, sadness, surprise, anticipation, anger, fear, trust, and disgust).

The PlutchikBERT tokenizes each post, and we use the embedding in the [CLS] as the representation of the emotional spectrum. Similarly, the emotion embeddings of the j-th SW and NSW can be computed as follows.

$$e_{\text{SW}_j}^{u_i} = \text{PlutchikBERT}(p_{\text{SW}_j}^{u_i}) \tag{3}$$

$$e_{\text{NSW}_j}^{u_i} = \text{PlutchikBERT}(p_{\text{NSW}_j}^{u_i}) \tag{4}$$

Therefore, the initialized embeddings of j-th SW and NSW post are:

$$x_{\text{SW}_j}^{u_i} = s_{\text{SW}_j}^{u_i} \oplus e_{\text{SW}_j}^{u_i}, \tag{5}$$

$$x_{\text{NSW}_j}^{u_i} = s_{\text{NSW}_j}^{u_i} \oplus e_{\text{SW}_j}^{u_i}, \tag{6}$$

where $\oplus$ represents the concatenation operation.

3.2   Bi-LSTM Layer

Given that Long-Short Term Memory network (LSTM) is proficient in capturing long term dependencies, we adopt the Bi-LSTM to encode the mental state features of a user over times. The specific formulas can be written as follows.

$$\overrightarrow{h_{\text{SW}_j}^{u_i}} = \overrightarrow{\text{LSTM}}\left(x_{\text{SW}_j}^{u_i}, \overrightarrow{h_{\text{SW}_{j-1}}^{u_i}}\right) \tag{7}$$

$$\overleftarrow{h_{\text{SW}_j}^{u_i}} = \overleftarrow{\text{LSTM}}\left(x_{\text{SW}_j}^{u_i}, \overleftarrow{h_{\text{SW}_{j+1}}^{u_i}}\right) \tag{8}$$

Then, the $\overrightarrow{h_{\text{SW}_j}^{u_i}}$ and $\overleftarrow{h_{\text{SW}_j}^{u_i}}$ are concatenate to represent the embedding of SW post $x_{\text{SW}_j}^{u_i}$ after Bi-LSTM, denoted as $h_{\text{SW}_j}^{u_i} = \overrightarrow{h_{\text{SW}_j}^{u_i}} \oplus \overleftarrow{h_{\text{SW}_j}^{u_i}}$. To embody diverse importance of SW posts, the following attention mechanism is adopted:

$$\tilde{\alpha}_j = c_r^{\text{T}}\tanh\left(W_r h_{\text{SW}_j}^{u_i} + b_r\right), \tag{9}$$

$$\alpha_j = \frac{\exp(\tilde{\alpha}_j)}{\sum_{j=1}^{n}\exp(\tilde{\alpha}_j)}, \tag{10}$$

where W_r is a weight matrix, b_r and c_r are bias terms. Finally, the output embeddings of SW posts are formulated as:

$$v_{\text{SW}}^{u_i} = \sum_{j=1}^{n}\alpha_j h_{\text{SW}_j}^{u_i}, \tag{11}$$

where $v_{\text{SW}}^{u_i} \in \mathbb{R}^{d_{sw}}$, d_{sw} is the dimension of $v_{\text{SW}}^{u_i}$.

3.3   T-LSTM Layer

Considering that users' mental states change over time in the real world, the interval of posting time should be encoded to simulate the changing process. Obviously, the long time interval between previous post and current post is less significant in revealing the psychological state of the user, while the weights between consecutive posts are identical in the existing LSTM model.

Following the literature [35], to capture the irregularity of posting time of historical posts, we develop a Time-Aware LSTM (T-LSTM), in which the time interval information between consecutive posts is incorporated into the LSTM. The working principle is shown in Fig. 2.

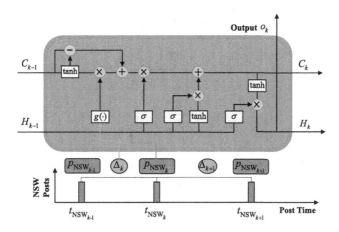

Fig. 2. Principle of Time-Aware LSTM cell.

In Fig. 2, the time decay information is added to the memory discount, weighing the short-term memory cell C_t^S; that is, the long-term memory does not be discarded entirely, and the significance of short-term information will be adjusted according to the elapsed time between two NSW posts at time steps $t_{\mathrm{NSW}_{t-1}}^{u_i}$ and $t_{\mathrm{NSW}_t}^{u_i}$. To model this idea, T-LSTM exploits a monotonically decreasing function to deal with the elapsed time and transforms it into weights. Mathematically, these operations can be defined as follows:

$$C_{t-1}^S = \tanh\left(W_k C_{t-1} + b_k\right), \tag{12}$$

$$\Delta_t = t_{\mathrm{NSW}_t}^{u_i} - t_{\mathrm{NSW}_{t-1}}^{u_i}, \tag{13}$$

$$g(\Delta_t) = 1/\Delta_t, \tag{14}$$

$$\hat{C}_{t-1}^S = C_{t-1}^S * g(\Delta_t), \tag{15}$$

$$C_{t-1}^{LT} = C_{t-1} - C_{t-1}^S, \tag{16}$$

$$C_{t-1}^* = C_{t-1}^{LT} + \hat{C}_{t-1}^S, \tag{17}$$

where C_{t-1} and C_t are previous and current cell memories, and W_k, b_k are the network parameters. C_{t-1}^S represents the short-term memory, Δ_t is the time interval between $t_{\mathrm{NSW}_{t-1}}^{u_i}$ and $t_{\mathrm{NSW}_t}^{u_i}$, $g(\cdot)$ is a decay function, $\hat{C}_{t-1}^S$ refers to the discounted short-term, C_{t-1}^{LT} represents the long-term memory and C_{t-1}^* is the adjusted previous cell memory.

For the NSW posts of user u_i, corresponding embeddings are poured into T-LSTM, along with an attention mechanism to learn adaptive weights for the hidden state of each post. Eventually, the final representation of all NSW posts is formed by summing their embeddings.

$$\left[h_{\mathrm{NSW}_1}^{u_i}, h_{\mathrm{NSW}_2}^{u_i}, \cdots, h_{\mathrm{NSW}_m}^{u_i}\right] = \text{T-LSTM}\left(x_{\mathrm{NSW}_1}^{u_i}, x_{\mathrm{NSW}_2}^{u_i}, \cdots, x_{\mathrm{NSW}_m}^{u_i}\right) \quad (18)$$

$$\tilde{\beta}_j = c_u^{\mathrm{T}} \tanh\left(W_u h_{\mathrm{NSW}_j}^{u_i} + b_u\right) \quad (19)$$

$$\beta_j = \frac{\exp\left(\tilde{\beta}_j\right)}{\sum_{j=1}^m \exp\left(\tilde{\beta}_j\right)} \quad (20)$$

$$v_{\mathrm{NSW}}^{u_i} = \sum_{j=1}^m \beta_j h_{\mathrm{NSW}_j}^{u_i} \quad (21)$$

W_u is a weight matrix, b_u and c_u are bias terms. $h_{\mathrm{NSW}_j}^{u_i} \in \mathbb{R}^{d_{nsw}}$, d_{nsw} is the dimension of each NSW post vector. $v_{\mathrm{NSW}}^{u_i} \in \mathbb{R}^{d_{nsw}}$.

3.4   User Features

In addition to the features in SuicideWatch (SW) and Non-Suicide Watch (NSW) posts, the LIWC (Linguistic Inquiry and Word Count) information is also valuable because it can detect individual's mental state [16,17]. Hence, we combine all words in SW posts of user u_i and use the LIWC dictionary to calculate the differences in words used of users.

We denote a 64-dimension vector $L^{u_i} = [l_1, l_2, \cdots, l_{64}]$ corresponding to the 64 word-categories of the LIWC dictionary, in which the value of each dimension represents the frequency of word belonging to the corresponding category in SW posts. The specific formula is: $l_k = word_k/word_{\mathrm{sw}}$, where l_k is the value of k-th dimension of L^{u_i}, $word_k$ represents the number of words belonging to k-th category, $word_{\mathrm{sw}}$ is the total number of words in SW posts.

Also, users' posting behaviors (e.g., posting period) can also demonstrate the differences in their psychological states [15]. To encode this information, a four-dimension vector $D^{u_i} = [d_1, d_2, d_3, d_4]$ is set, where $d_k = N_k/N$ represents the proportions of all user's posts posted in the period (0:00–5:59), (6:00–11:59), (12:00–17:59), (18:00–23:59) of a day; N_k refers to the number of posts posting in k-th period of a day, and N is the total number of posts of user u_i.

Table 1. Statistics of users and posts on the Reddit

Labels	Train			Test		
	Users	SW Posts	NSW Posts	Users	SW Posts	NSW Posts
NR	127(25.6%)	162	10,500	32(25.6%)	34	2,616
LR	50(10.1%)	59	2,656	13(10.4%)	13	533
MR	113(22.8%)	154	5,572	28(22.4%)	41	2,194
SR	206(41.5%)	544	11,906	52(41.6%)	98	4,081
Total	496	919	30,634	125	186	9,424

3.5 Classification Layer

After the above processes, for the user u_i, we attain the embeddings of SW posts $(v_{SW}^{u_i})$, user mental state changes in NSW posts $(v_{NSW}^{u_i})$, LIWC feature (L^{u_i}), and posting time (D^{u_i}) of posts. Then, we concatenate the four embeddings and feed them into a fully-connection network, followed by a softmax function to compute the distribution over four suicide risk levels:

$$\hat{p} = \text{softmax}(W_t(\text{UE}) + b_t), \tag{22}$$

where $\text{UE} = v_{SW}^{u_i} \oplus v_{NSW}^{u_i} \oplus L^{u_i} \oplus D^{u_i}$, W_t denotes a weight matrix, and b_t is a bias term.

4 Experiment

4.1 Dataset and Evaluation

Following previous work [34], we selected the corpus *Reddit* published for the subtask B in the CLPsych2019 Shared Task as the experimental dataset, which consists of SuicideWatch and Non-SuicideWatch posts. Experts and crowdsource workers annotate the suicide risk level for each user with one of the following four levels, including No Risk (NR), Low Risk (LR), Moderate Risk (MR), and Severe Risk (SR). Table 1 reports the statistics of users and posts for each suicide risk level in Reddit.

We employ the official metrics for CLPsych2019 shared tasks (i.e., Macro-$F1$, accuracy (Acc), SR-$F1$ and MR-$F1$) in all evaluations.

4.2 Baselines and Hyper-Parameter

To comprehensively evaluate our SRED model, we compare it with the best models in CLPsych2019 task A and the top three models in task B that both exploit SW posts and NSW posts.

- **CLaC** [10] used the GloVe and ELMo to encode posts and input corresponding embeddings into CNN, BI-RNN, BI-LSTM and BI-GRU to learn features, followed by the SVM serving as the classifier, respectively.

- **SBU-HLAB** [11] adopted BERT to encoded SW and NSW posts separately.
- **CAMH**[1] is the second place on CLPsych2019 task B [34].
- **TsuiLab** [36] created an ensemble model based on the top three models (e.g., NB, SVM, Gradient Boosting).

In addition to the best models in CLPsych2019 task, we also compared with state-of-the-art models.

- **SDM** [28] encoded posts with fine-tuned FastText embeddings, and posts were passed sequentially through LSTM with attention.
- **STATENet** [32] exploited the fine-tuned Transformer for post encoding, and the embeddings were fed into a T-LSTM. The hidden state of the latest post represents the user's historical emotion context.
- **PHASE** [7] employed a fine-tuned Transformer to encode posts and learned users' emotional phase progressions from historical emotion context.

Given that SDM, STATENet and PHASE detected suicide ideation on posts-level and did not divide suicidal posts and other posts, we compared with two different strategies. One is to implement the models on SW and NSW posts, respectively; the other is to merge SW posts and NSW posts in chronological order.

Regarding hyper-parameters, we choose the AdamW optimizer, and the hidden state dimension of Bi-LSTM and T-LSTM is set to 256. The number of LSTM layers is 1. Also, the learning rate, dropout, and warm-up steps are set to 1e-3, 0.5, and 5, respectively. The batch size and epoch are set to 128 and 70. The experimental environment of this paper is PyTorch 1.7 and Nvidia Geforce RTX 2080 Ti GPU.

4.3   Overall Performance

Table 2 reports the experimental results, where "_S" denotes that the SW and NSW posts are modeled separately, "_C" indicates that the SW and NSW posts are combined in chronological order. Bold font denotes the best performance.

It can be seen that our model outperforms the top three models of task B in CLPsych2019. Specifically, the Macro-$F1$, LR-$F1$ and MR-$F1$ of SRED are superior to that of SBU-HLAB (the first place of task B) with 4.9%, 13.5% and 6.6%, suggesting the advantages of using emotion traits in posts. Also, compared with the CLaC baseline, SRED achieves 10.7% and 14.1% improvement in SR-$F1$ and LR-$F1$, respectively. It confirms that the emotional changes in the NSW posts can improve the model's ability to recognize the specific level of user suicide risk, especially SR and LR labels.

Finally, looking at the baselines of SDM (lines 5–6), STATENet (lines 7–8) and PHASE (lines 9–10), our SRED is also sophisticated whether encoding the mixed posts (SW and NSW) or separately modeling them. This is because SRED

---

[1] No paper is available and the official task web does not demonstrate the implementation details.

Table 2. Experimental results of different models

Models	Acc	Macro-$F1$	SR-$F1$	MR-$F1$	LR-$F1$	NR-$F1$
CLaC	50.4	48.1	54.3	**40.0**	24.4	**73.7**
SBU-HLAB	56.0	45.7	**69.9**	24.5	25.0	63.4
CAMH	51.2	41.3	59.8	22.6	10.5	72.1
TsuiLab	40.8	37.0	50.6	26.4	20.5	50.7
SDM_S	47.2	38.4	61.2	14.0	18.1	60.0
SDM_C	37.6	26.0	51.8	17.6	11.0	23.7
STATENet_S	48.0	40.5	61.5	17.1	35.9	47.5
STATENet_C	44.0	27.3	59.1	17.6	13.3	19.0
PHASE_S	48.8	41.8	61.8	27.9	23.5	54.0
PHASE_C	36.0	27.5	52.0	14.6	18.2	25.0
SRED(Ours)	**56.8**	**50.6**	65.0	31.1	**38.5**	67.7

adapts the encoding strategy according to the diverse characteristics of posts. Also, since the information on SW posts is mixed with some noise, it is difficult to identify user-level suicide risk if users' historical posts are encoded directly without dividing them into SW and NSW posts.

5 Additional Analyses

To further investigate the specific impact of each component and increase the interpretability of the suicide risk assessment performance, we conducted additional experiments, including ablation study and qualitative analysis cases.

5.1 Ablation Study

To verify the effectiveness of T-LSTM, emotion embedding and user features, we conducted an ablation study with different configurations, and Table 3 reports the results.

As shown in Table 3, without NSW posts (line 2), the Macro-$F1$ decreases by 7.3% and NR-$F1$ has a sharp drops, demonstrating that NSW posts are meaningful for predicting the suicide risk level by providing the information that reveals user's mental health states. When replacing the T-LSTM with Bi-LSTM to model NSW posts (line 3), the Macro-$F1$, MR-$F1$ and NR-$F1$ drop drastically, suggesting a more significant impact of time features on the final performance.

Deactivating the emotion features (line 4) results in an 8.1% points degradation in Macro-$F1$. If removing the user features (i.e., LIWC word frequency and posting time), the MR-$F1$ and LR-$F1$ decline, confirming the effectiveness of user features in identifying users of MR and LR.

Table 3. Results of ablation study over SRED

No.	Models	Acc	Macro-$F1$	SR-$F1$	MR-$F1$	LR-$F1$	NR-$F1$
1	SRED	**56.8**	**50.6**	65.0	**31.1**	**38.5**	67.7
2	w/o NSW	50.4	43.3	67.3	26.9	23.5	55.6
3	w/o Time	54.4	45.6	**68.4**	24.2	34.3	55.4
4	w/o Emotion	48.0	42.5	66.0	26.7	22.2	54.9
5	w/o User Features	56.8	49.1	65.0	27.3	33.3	**71.0**

5.2   SRED Qualitative Analysis

For a detailed insight and aiding interpretability, we analyze three user cases, in which SRED performs well. We fed posts from the three users into PlutchikBERT and obtained probability distributions of eight primary emotions for each post. Figure 3 and Fig. 4 demonstrate the emotional probability distribution of SW posts and NSW posts over time of User A to C, respectively.

It can be seen that partial users just ask friends for advice in SW. For instance, when the word "feeling suicidal" appears in the SW post of User A, it is a sign of suicidal intent, while the one really having the suicidal ideation is a friend of User A. Along this line, User A may be easily mistaken as at suicide risk by merely analyzing SW posts. However, User A demonstrates the positive emotion characterized by joy (Fig. 4a) and does not express any suicidal thoughts in NSW posts.

Also, when there are no explicit suicidal signs in user SW posts, NSW posts can improve the identification performance of suicide ideation by adding temporal emotional information. In Fig. 3b, although the SW post demonstrates the negative emotion, User B is not indicative of suicidal intent. However, SRED can learn the build-up of sadness in Fig. 4b, resulting in the correct recognition.

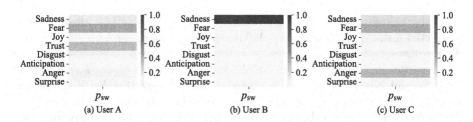

Fig. 3. The fine-grained emotional distribution in SW post of User A to C.

Furthermore, given explicit suicidal signs in some SW posts, further support from NSW posts is often needed to classify users into specific risk levels. For example, User C shows a suicide attempt in the SW post, while it is sticky to identify his suicide risk level. SRED aggregates complex emotions of sadness, fear

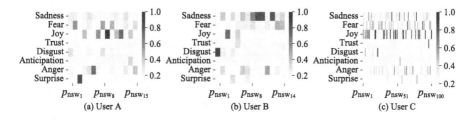

Fig. 4. The fine-grained emotional distribution in NSW posts of User A to C.

and joy (the dominant emotion) and assumes that User C's suicidal attempt is not very urgent. Therefore, it is reasonable to identify User C as MR instead of SR.

6   Conclusion

This paper aims to assess the suicide risk levels of users. Given that the Non-SuicideWatch data contains essential information for understanding their mental states and the emotional changes within a period of time, we propose a Suicide Risk Detection model (SRED) based on the dual context (i.e., SuicideWatch and Non-SuicideWatch). First, the suicidal context from SuicideWatch posts is exploited. Second, the temporal emotional changes of Non-SuicideWatch posts are explored to understand users' mental states, whereby improving the identification of the specific severity level of users' suicide risk. Compared with the best models in the CLPsych2019 shared task and the current advanced models, SRED demonstrates advancement and effectiveness. We also present a qualitative analysis to increase the interpretability of SRED. Through SRED, we can provide a preliminary screening of different suicide risk levels of users to aid the prioritization of medical resources.

Acknowledgements. This work is supported by the Natural Science Foundation of China under Grants 62272206, 62272205, 61972184, and 62076112; Foundation for Academic and Technical Leaders in Major Disciplines of Jiangxi Province Grant 2021BCJL22041; and Natural Science Foundation of Jiangxi Province Grant 20212ACB202002. Additionally, we would like to thank the reviewers for their feedback and suggestions on revisions.

References

1. WHO: Suicide worldwide in 2019: global health estimates. World Health Organization (2021)
2. Holly, H., Curtin, S.C., Margaret, W.: Increase in suicide mortality in the united states. In: NCHS Data Brief, vol. 362, pp. 1–8 (2020)
3. Sawhney, R., Joshi, H., Gandhi, S., Jin, D., Shah, R.R.: Robust suicide risk assessment on social media via deep adversarial learning. J. Am. Med. Inform. Assoc. **28**(7), 1497–1506 (2021)

4. Skaik, R., Inkpen, D.: Using social media for mental health surveillance: a review. ACM Comput. Surv. (CSUR) **53**(6), 1–31 (2020)
5. Mathur, P., Sawhney, R., Chopra, S., Leekha, M., Ratn Shah, R.: Utilizing temporal psycholinguistic cues for suicidal intent estimation. In: European Conference on Information Retrieval, pp. 265–271 (2020)
6. Sinha, P.P., Mishra, R., Sawhney, R., Mahata, D., Shah, R.R., Liu, H.: # suicidal-A multipronged approach to identify and explore suicidal ideation in twitter. In: Proceedings of the 28th ACM International Conference on Information and Knowledge Management, pp. 941–950 (2019)
7. Sawhney, R., Joshi, H., Flek, L., Shah, R.R.: PHASE: learning emotional phase-aware representations for suicide ideation detection on social media. In: Proceedings of the 16th Conference of the European Chapter of the Association for Computational Linguistics: Main Volume, pp. 2415–2428 (2021)
8. Gaur, M., et al.: Knowledge-aware assessment of severity of suicide risk for early intervention. In: The World Wide Web Conference, pp. 514–525 (2019)
9. Ren, F., Kang, X., Quan, C.: Examining accumulated emotional traits in suicide blogs with an emotion topic model. IEEE J. Biomed. Health Inform. **20**(5), 1384–1396 (2016)
10. Mohammadi, E., Amini, H., Kosseim, L.: CLaC at CLPsych 2019: fusion of neural features and predicted class probabilities for suicide risk assessment based on online posts. In: Proceedings of the Sixth Workshop on Computational Linguistics and Clinical Psychology, pp. 34–38 (2019)
11. Matero, M., et al.: Suicide risk assessment with multi-level dual-context language and BERT. In: Proceedings of the Sixth Workshop on Computational Linguistics and Clinical Psychology, pp. 39–44 (2019)
12. Devlin, J., Chang, M.-W., Lee, K., Toutanova, K.: BERT: pre-training of deep bidirectional transformers for language understanding. In: Proceedings of the 2019 Conference of the North American Chapter of the Association for Computational Linguistics: Human Language Technologies, vol. 1, pp. 4171–4186 (2019)
13. Plutchik, R.: A general psychoevolutionary theory of emotion. In: Theories of Emotion, pp. 3–33 (1980)
14. Abdul-Mageed, M., Ungar, L.: EmoNet: fine-grained emotion detection with gated recurrent neural networks. In: Proceedings of the 55th Annual Meeting of the Association for Computational Linguistics (Volume 1: Long Papers), pp. 718–728 (2017)
15. Peng, Z., Hu, Q., Dang, J.: Multi-kernel SVM based depression recognition using social media data. Int. J. Mach. Learn. Cybern. **10**(1), 43–57 (2019)
16. De Choudhury, M., Kiciman, E., Dredze, M., Coppersmith, G., Kumar, M.: Discovering shifts to suicidal ideation from mental health content in social media. In: Proceedings of the 2016 CHI Conference on Human Factors in Computing Systems, pp. 2098–2110 (2016)
17. Guan, L., Hao, B., Cheng, Q., Yip, P.S., Zhu, T.: Identifying Chinese microblog users with high suicide probability using internet-based profile and linguistic features: classification model. JMIR Mental Health **2**(2), 17 (2015)
18. Tsugawa, S., Kikuchi, Y., Kishino, F., Nakajima, K., Itoh, Y., Ohsaki, H.: Recognizing depression from twitter activity. In: Proceedings of the 33rd Annual ACM Conference on Human Factors in Computing Systems, pp. 3187–3196 (2015)
19. Masuda, N., Kurahashi, I., Onari, H.: Suicide ideation of individuals in online social networks. PLoS ONE **8**(4), 62262 (2013)

20. De Choudhury, M., Gamon, M., Counts, S., Horvitz, E.: Predicting depression via social media. In: Seventh International AAAI Conference on Weblogs and Social Media (2013)
21. Braithwaite, S.R., Giraud-Carrier, C., West, J., Barnes, M.D., Hanson, C.L.: Validating machine learning algorithms for twitter data against established measures of suicidality. JMIR Mental Health **3**(2), 21 (2016)
22. Ji, S., Yu, C.P., Fung, S.-F., Pan, S., Long, G.: Supervised learning for suicidal ideation detection in online user content. Complexity **2018**, 1–10 (2018)
23. Lin, H., et al.: Detecting stress based on social interactions in social networks. IEEE Trans. Knowl. Data Eng. **29**(9), 1820–1833 (2017)
24. Trotzek, M., Koitka, S., Friedrich, C.M.: Utilizing neural networks and linguistic metadata for early detection of depression indications in text sequences. IEEE Trans. Knowl. Data Eng. **32**(3), 588–601 (2020)
25. Husseini Orabi, A., Buddhitha, P., Husseini Orabi, M., Inkpen, D.: Deep learning for depression detection of Twitter users. In: Proceedings of the Fifth Workshop on Computational Linguistics and Clinical Psychology: From Keyboard to Clinic, pp. 88–97 (2018)
26. Sawhney, R., Manchanda, P., Singh, R., Aggarwal, S.: A computational approach to feature extraction for identification of suicidal ideation in tweets. In: Proceedings of ACL 2018, Student Research Workshop, pp. 91–98 (2018)
27. Yates, A., Cohan, A., Goharian, N.: Depression and self-harm risk assessment in online forums. In: Proceedings of the 2017 Conference on Empirical Methods in Natural Language Processing, pp. 2968–2978 (2017)
28. Cao, L., et al.: Latent suicide risk detection on microblog via suicide-oriented word embeddings and layered attention. In: Proceedings of the 2019 Conference on Empirical Methods in Natural Language Processing and the 9th International Joint Conference on Natural Language Processing, pp. 1718–1728 (2019)
29. Shing, H.-C., Resnik, P., Oard, D.: A prioritization model for suicidality risk assessment. In: Proceedings of the 58th Annual Meeting of the Association for Computational Linguistics, pp. 8124–8137 (2020)
30. Cao, L., Zhang, H., Wang, X., Feng, L.: Learning users inner thoughts and emotion changes for social media based suicide risk detection. IEEE Trans. Affect. Comput., 1 (2021)
31. Sawhney, R., Joshi, H., Shah, R.R., Flek, L.: Suicide ideation detection via social and temporal user representations using hyperbolic learning. In: Proceedings of the 2021 Conference of the North American Chapter of the Association for Computational Linguistics: Human Language Technologies, pp. 2176–2190 (2021)
32. Sawhney, R., Joshi, H., Gandhi, S., Shah, R.R.: A time-aware transformer based model for suicide ideation detection on social media. In: Proceedings of the 2020 Conference on Empirical Methods in Natural Language Processing, pp. 7685–7697 (2020)
33. Sawhney, R., Joshi, H., Gandhi, S., Shah, R.R.: Towards ordinal suicide ideation detection on social media. In: Proceedings of the 14th ACM International Conference on Web Search and Data Mining, pp. 22–30 (2021)
34. Zirikly, A., Resnik, P., Uzuner, O., Hollingshead, K.: CLPsych2019 shared task: predicting the degree of suicide risk in Reddit posts. In: Proceedings of the Sixth Workshop on Computational Linguistics and Clinical Psychology, pp. 24–33 (2019)

35. Baytas, I.M., Xiao, C., Zhang, X., Wang, F., Jain, A.K., Zhou, J.: Patient subtyping via time-aware LSTM networks. In: Proceedings of the 23rd ACM SIGKDD International Conference on Knowledge Discovery and Data Mining, pp. 65–74 (2017)
36. Ruiz, V., et al.: CLPsych2019 shared task: predicting suicide risk level from Reddit posts on multiple forums. In: Proceedings of the Sixth Workshop on Computational Linguistics and Clinical Psychology, pp. 162–166 (2019)

Item Recommendation on Shared Accounts Through User Identification

Chongming Gao[1]([✉]) [ID], Min Wang[2] [ID], and Jiajia Chen[1] [ID]

[1] University of Science and Technology of China, Hefei, China
chongming.gao@gmail.com
[2] Anhui Medical University, Hefei, China

Abstract. Nowadays, people often share their subscription accounts, e.g. online content subscription accounts, among family members and friends. It is important to identify different users under one single account and then recommend specific items to decoupled individuals. In this paper, we propose the Projected Discriminant Attentive Embedding (PDAE) model and the Shared Account-aware Bayesian Personalized Ranking (SA-BPR) model for user identification and item recommendation, respectively. PDAE separates item consumption actions of each individual from mixed account history by learning the item representation that has both the user preference and user demographic information integrated; SA-BPR is a robust recommendation model based on a hierarchical ranking strategy where items from different sets are recommended with different levels of priorities. The experiments show that the proposed models generally outperform state-of-the-art approaches in both user identification and item recommendation.

Keywords: User Identification · Recommendation · Shared Accounts

1 Introduction

Nowadays, sharing accounts of some online services becomes pervasive among Internet users. It is natural for friends or family members to share accounts or devices in some cases, such as watching the smart TV or speaking to a smart speaker, e.g., Google Home, and sharing a premium account to enjoy online streaming services provided by YouTube. In these cases, Recommender Systems (RS) play an important role in satisfying users by providing content that the users may love.

However, traditional recommendation methods [5–10] assume that one account is owned by one user, which impairs the performance of recommendations, since every member has his/her own preferences. To solve the problem, we propose to identify multiple users behind the shared accounts, and then recommend items according to the identification result. More specifically, there are three main steps to proceed: First, given an account and its historical logs, we attempt to detect a group of users and ascribe the account activities to the

F. Wu et al. (Eds.): SMP 2023, CCIS 1945, pp. 63–76, 2024.
https://doi.org/10.1007/978-981-99-7596-9_5

right issuer. Second, when a user starts to use the account and generates new session data, we can assign this new session to a specific user among the recognized users. Third, we aim to boost the performance of recommendations by user identification.

Although user identification has been studied in previous studies, most of them solved it either by implicitly modeling user preferences in finer granularity [1,28,29,32,33,36], or by proposing session-based methods in a short period to circumvent multiple-users cases [11,12,19,26]. There are only a few studies [16, 30,35] that directly focused on the problem of users identification. Among these works, White et al. [30] used a supervised manner to directly learn the mapping between users and accounts, which is not practical since the ground truth cannot be obtained in most situations. Zhang et al. [35] attempted to identify users by applying subspace clustering on user-item interaction histories while neglecting all other side information. Furthermore, Jiang et al. [16] distinguished users by learning the item embedding on the item meta graph.

The basic logic behind these studies is that users behind a shared account can be distinguished by their preferences. For example, one who appreciates the beauty of classical music can easily be distinguished from another who prefers Rock. However, none of the studies considered explicitly modeling user information, which directly affects the identification performance.

Therefore, we lay emphasis on the importance of user profiles, i.e. user demographic information such as gender, age, and country that no one leveraged before. As discovered in [4], statistically, people of different ages or genders have different preferences. Therefore, we can integrate user demographic information into item representations. In this paper, we propose two models for user identification and item recommendation respectively.

User Identification. We propose the **P**rojected **D**iscriminant **A**ttentive **E**mbedding (PDAE) model. PDAE firstly generates random walk sequences on both the user-item interaction graph and item meta graph, and it adopts the skip-gram architecture to capture the basic user preferences [22]. At the same time, PDAE integrates the user demographic information by projecting the users' attentive features to the user profile spaces and distinguishing the users with the same profiles (e.g., age, gender) from those with different ones.

Item Recommendation. We propose the **S**hared **A**ccount-aware **B**ayesian **P**ersonalized **R**anking (SA-BPR) model. When recommending items to the identified user, we have to consider the situation where the user is mistakenly identified. SA-BPR puts forward a remedy by letting a user also prefer the items consumed by others in the account, and the intensity of preference is determined by the confidence of identification results. Specifically, for a user in a shared account, there are three kinds of items: $\mathcal{I}_u$: items consumed by this user, $\mathcal{I}_o$: items consumed by others in the same account, and $\mathcal{I}_n$: items never consumed by anyone in this account, i.e., negative samples. SA-BPR assumes that the importance (rankings) of the three kinds of items satisfy: $\mathcal{I}_u > \mathcal{I}_o > \mathcal{I}_n$.

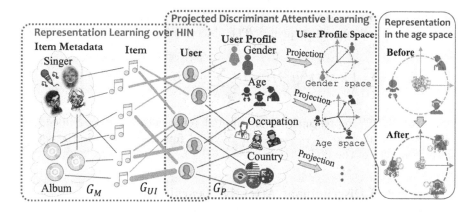

Fig. 1. Representation learning for user identification.

2 Problem Formulation

Let $\mathcal{I}$ be the set of items, e.g., songs. For each item $i \in \mathcal{I}$, it may have multiple attributes, e.g., artist(s) and genre(s) of a song. We denote the set $\mathcal{M}_i$, as metadata of item i. All items and their attributes compose an item meta-network, denoted as G_M. Let $\mathcal{A}$ be the set of all accounts. Each $a \in \mathcal{A}$ contains a set of users, denoted as $\mathcal{U}_a$. We denote the user-item bipartite graph as G_{UI}. We combine G_M and G_{UI} together as G. For each user $u \in \mathcal{U}$, it may contain some profile attributes, e.g., age, gender, country. We denote the collection of all profile attributes as $\mathcal{P}$ and the network contained all users and profile attributes as G_P. For computational efficiency, we split the values of each profile attribute $p \in \mathcal{P}$ into several categories, i.e., $p = \{c_1, c_2, \cdots, c_{|p|}\}$. For example, all age values are split into three categories: Children, Adults, and the Elderly. $f(x) \in \mathbb{R}^{(n)}$ denotes the n-dimensional feature vector of the variable x.

3 User Identification

3.1 Heterogeneous Graph Embedding

Considering that the key to identifying users is to capture their preferences according to the items they consumed, we propose our representation learning model over the heterogeneous Information Network (HIN). The HIN G includes both the user-item interaction graph G_{UI} and the item meta graph G_M, as illustrated in Fig. 1. To simplify the problem, we extend the basic skip-gram architecture [22] to capture the network structure. In natural language processing (NLP), the skip-gram architecture learns relations between words and their context in sentences. In our work, each node on the HIN is treated as a word, and sentences are simulated by random walking over the HIN. For node v on the G, the negative log-likelihood of its contexts $C(v)$ conditioned on v is minimized.

The contexts $C(v)$ are selected as nodes within a window that is centered on node v on the random walk sequence. The objective function is written as:

$$O_h = -\sum_{v \in G} \left(\log \sigma(f(v)^T \cdot f(v_c)) + \sum_{v_n \in N_s(v)} \log \sigma(-f(v)^T \cdot f(v_n)) \right), \qquad (1)$$

where $f(v) \in \mathbb{R}^{(n)}$ represents the distributed representation (i.e., embedding vector) of node v. $\sigma(x) = \frac{1}{1+e^{-x}}$ is the Sigmoid function. $N_s(v)$ is the set of negative samples of the node v.

3.2   Projected Discriminant Attentive Learning

As we have discovered that users with different demographic profiles are likely to have different preferences, it is vital to integrate the information into item representation.

Projected Attentive Embedding. For a user u, the representation of it in a space related to the profile $p \in \mathcal{P}$ is denoted as the feature vector: $f_p(u) \in \mathbb{R}^{(m)}$. The relationship between u's heterogeneous embedding vector $f(u) \in \mathbb{R}^{(n)}$ and all of u's profile vectors in all profile spaces $\{f_p(u), p \in \mathcal{P}\}$ is one-to-many. To let the user demographics explicitly influence items, we consider projecting items into the user profile space. Since items do not possess user demographic attributes, we make the user automatically pay "attention" to items that they love and automatically produce the projected item embedding vectors. Inspired by the graph attention network (GAT) architecture [27], we build a single graph attention layer for users on the graph G_{UI}. Specifically, for user u on the graph G_{UI}, suppose all items he/she consumed are denoted as set $\mathcal{I}_u = \{i_1, i_2, \cdots, i_{|\mathcal{I}_u|}\}$. For each item $i \in \mathcal{I}_u$, the normalized attention coefficient is define as:

$$\alpha_{ui} = \frac{\exp\left(\text{LeakyReLU}\left(\vec{a}^T[\mathbf{W} \cdot f(u) || \mathbf{W} \cdot f(i)]\right)\right)}{\sum_{j \in \mathcal{I}_u} \exp\left(\text{LeakyReLU}\left(\vec{a}^T[\mathbf{W} \cdot f(u) || \mathbf{W} \cdot f(j)]\right)\right)},$$

where $||$ is the concatenation operation. $\mathbf{W} \in \mathbb{R}^{(n \times n)}$ is a weight matrix applied to every node. $\vec{a} \in \mathbb{R}^{(n)}$ is a weight vector. We set the activation function as LeakyReLU [20].

The normalized attention coefficient α_{ui} is used to compute the attentive feature of user u as: $f'(u) = \sigma\left(\sum_{i \in \mathcal{I}_u} \alpha_{ui} \mathbf{W} \cdot f(i)\right)$. We combine the original and attentive user representation as:

$$f_p(u) = \mathbf{A}_p \cdot (f(u) + f'(u)), \qquad (2)$$

Discriminant Learning in Profile Space. After obtaining the feature $f_p(u)$ for every user u who possesses profile attribute p, we separate these features according to the profile categories. For example, the representation of children, adults, and the elderly should be classified into different clusters. These clusters should be pushed to locate as far as possible from each other for identification.

Projected Discriminant Attentive Embedding (PDAE). For each profile category $c_i \in p$, e.g., *Children* in the profile attribute *Age*, the representation should be kept away from other profile categories $c_j \in p, (j \neq i)$, e.g., *Adult*. We formulate the objective function as the pairwise loss between every category pair:

$$O_p = \sum_{p \in \mathcal{P}} \sum_{c_i \in p} \sum_{c_j \in p, j > i} \log \sigma(f(c_i)^T \cdot f(c_j)), c \in p : \|f(c)\|_F \leq 1, \qquad (3)$$

where $f(c)$ is the embedding vector of profile category c, and it is restricted in the unit sphere to avoid the trivial solution. Then the objective function for all users is:

$$O_d = - \sum_{p \in \mathcal{P}} \sum_{c \in p} \sum_{u \in \mathcal{U}_c^p} \log \sigma(f(c)^T \cdot f_p(u)), \qquad (4)$$

where $\mathcal{U}_c^p$ is the set of users that have profile attribute p and the profile value is category c. $f_p(u)$ is the projected user feature vector defined in Eq. (2).

By combining the heterogeneous graph embedding learning and the projected attentive embedding learning as well as the discriminant learning, we get the objective function of PDAE as:

$$O_{PDAE} = O_h + O_p + O_d. \qquad (5)$$

User Identification and Session Assignment. Given all historical sessions $\mathcal{S}_a$ of a shared account a, we obtain the features of sessions. we want to detect a set of users behind this account and ascribe sessions to their actual issuers. However, the number of users N is unknown. It's not practical to automatically estimate N according to the learned representation because (1) A user can have multiple preferences and it is liable to be recognized as many individuals. Actually, it is inevitable. (2) Different users sharing the same account may have preferences overlapped and they can be identified as one person. Thereby, we manually assign N and use the well-known k-means algorithm to find N users, i.e., *virtual personalities*, in an account. We will demonstrate in the experiment section that even if there are misidentified cases in user identification, The performance of item recommendation is still satisfactory because the preferences of the identified virtual personalities have been captured and modeled.

Once the N users and their corresponding sessions are obtained, The recommendation model (described later) is built for N users separately. When there is a new session started in the account, the system will predict which user, among the N users, is issuing this session and recommend items according to the identified user. To achieve this goal, we apply the k-nearest neighbors (k-NN) algorithm for the new-coming session s, where K nearest historical session feature vectors are found and the new session is assigned to the user who has most of the sessions in the K neighbors.

4 Recommendation for Identified User

4.1 SA-BPR: Shared Account-Aware BPR

Different from traditional recommender systems where the users are considered independent of each other, the identified users (*virtual personalities*) are not precisely corresponded to the ground truth users. For example, if a user is identified as having multiple personalities, the recommendation model built on one of his/her personalities can not cover his/her preferences.

To overcome this problem, we build the recommendation system not only for the identified user but also for others in the same accounts. we extend the powerful Bayesian Personalized Ranking (BPR) model [24] to work properly in the account sharing situation. Formally, given an account a along with its N identified users $\mathcal{U}_a$, for an identified user $u \in \mathcal{U}_a$, we define three types of items according to their relationship with user u:

- $\mathcal{I}_u$: items consumed by user u;
- $\mathcal{I}_o$: items consumed by others in the same account;
- $\mathcal{I}_n$: items never consumed by this account, i.e., negative samples.

Note that $\mathcal{I}_u \cup \mathcal{I}_o \cup \mathcal{I}_n = \mathcal{I}$ and they are disjoint with each other. SA-BPR assumes the preference of user $u \in \mathcal{U}_a$ towards the three set of items is: $\mathcal{I}_u \succeq \mathcal{I}_o \succeq \mathcal{I}_n$, which derives the following relations:

$$x_{ui} \succeq x_{uo}, x_{uo} \succeq x_{un}, i \in \mathcal{I}_u, o \in \mathcal{I}_o, n \in \mathcal{I}_n.$$

where x_{ui} denotes the preference score of user u on item i. The preference function is modeled by matrix factorization:

$$x_{ui} = \mathbf{Z}_u^T \mathbf{Q}_i + \mathbf{b}_i, \tag{6}$$

where $\mathbf{Z} \in \mathbb{R}^{(d \times |\mathcal{U}|)}, \mathbf{Q} \in \mathbb{R}^{(d \times |\mathcal{I}|)}$ denotes the d-dimensional latent feature vectors of all users $\mathcal{U}$ and items $\mathcal{I}$, respectively; $\mathbf{b} \in \mathbb{R}^{(|\mathcal{I}|)}$ represents the bias vectors for items. We minimize the following objective function.

$$\min_{\mathbf{Z},\mathbf{Q},\mathbf{b}} -\sum_{a\in\mathcal{A}}\sum_{u\in\mathcal{U}_a}\left[\sum_{i\in\mathcal{I}_u}\sum_{o\in\mathcal{I}_o}\log\sigma(\frac{x_{ui}-x_{uo}}{1+g_u})+\right.$$
$$\left.\sum_{o\in\mathcal{I}_o}\sum_{n\in\mathcal{I}_n}\log\sigma(x_{uo}-x_{un})\right]+\frac{\lambda}{2}\left(||\mathbf{Z}||_F^2+||\mathbf{Q}||_F^2+||\mathbf{b}||_F^2\right), \tag{7}$$

where λ is a regularization coefficient, which is set to 0.01 in this work; g_u is the parameter to control how much the model would penalize the relation $x_{ui} \succeq x_{uo}$ compared to $x_{uo} \succeq x_{un}$. A large value of g_i would shrink the preference gap between $\mathcal{I}_u$ and $\mathcal{I}_o$, i.e., the user u would have a similar preference for items consumed by other sharers to the same extent as for his/her own consumed items. We use stochastic gradient descent (SGD) to optimize Eq. (7). Therefore, if g_i grows large enough, Eq. (7) will abandon differentiating users in accounts, and the problem will degenerate to the account-level item recommendation by using ordinary BPR model [24]. For computational efficiency, in this work, we optimize Eq. (7) by sampling 3 items in $\mathcal{I}_o$ and 3 items in $\mathcal{I}_n$ for each $u \in \mathcal{U}_a$.

Table 1. Information of three real-world datasets

Dataset		Movie	Book	Music
Basic Info Original / Processed	#**Users**	6040 / 6040	105283 / 13097	992 / 985
	#**Items**	3706 / 3706	340556 / 11327	1083481 / 12884
	#**Items per user**	165.6 / 165.6	10.9 / 33.29	4443.5 / 1261.3
	#**records per user**	165.6 / 165.6	10.9 / 33.29	19252.9 / 7067.7
User **Demographics** # types (% of valid users)	**Age**	7 (100%)	3 (59%)	3 (29%)
	Gender	2 (100%)	✗	2 (89%)
	Country	✗	5 (81%)	5 (91%)
	Occupation	20 (100%)	✗	✗
Item Metadata		#Genre:18	#Author:101608	#Artiest:176948

4.2   Adaptive Preference Adjustment

There is still a problem existing in SA-BPR: How to choose g_i to adjust the optimization intensity for preferences of the three types of items? The intuition is that we should widen the gap between the preferences for $\mathcal{I}_u$ and $\mathcal{I}_o$ (i.e., decrease g_u) for an account if the following conditions are met:

1. The number of identified users N is large.
2. The session feature vectors of user u are compact, i.e., close to the cluster center.
3. Cluster centers are distant from each other.

Therefore, we propose to set g_u adaptively as:

$$g_u = \frac{\text{Var}(u)}{\text{BS}(a)} = \frac{\left(\sum_{k=1}^{N_u} ||f(s_k) - \vec{\mu}_u||_F^2\right)/N_u}{\left(\sum_{1 \le i < j \le N} ||\vec{\mu}_i - \vec{\mu}_j||_F^2\right)/\frac{N*(N-1)}{2}}, \tag{8}$$

where $\text{Var}(u)$ is the variance of session feature vectors for user u; $\text{BS}(a)$ is the Between-class Scatter, which is defined as the average distance of all cluster pairs; N is the number of users; N_u is the number of sessions u issued; $f(s_k)$ is the feature vector of session s_k. $\vec{\mu}_u = \left(\sum_{k=1}^{N_u} f(s_k)\right)/N_u$ is the average of all session feature vectors of user u.

By plugging Eq. (8) in the objective function Eq. (7), SA-BPR can learn automatically from the results of user identification. If the user identification for an account is of high confidence, i.e., the three conditions above are satisfied, the recommender system would be tailored to each identified user behind the account rather than the account as a whole.

4.3   Recommendation

SA-BPR model is trained on the historical user identification results. When an account issues a new session, it can be assigned to one of the identified users, e.g., by k-NN algorithm. Then for the identified user, we can recommend items by calculating Eq. (6) and selecting items that have the maximum values of x_{ui}.

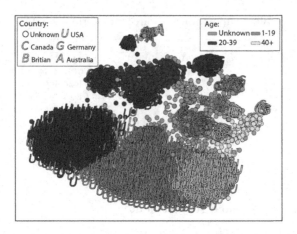

Fig. 2. t-SNE results of user representations.

5 Experimental Evaluation

In this section, we evaluate the proposed models on three real-world scenarios: music, movie, and book. We have two evaluation tasks: (1) evaluating user identification on both the historical data and new data; (2) evaluating the individual-level recommendation according to the identification results.

5.1 Experimental Setup

Data Preparation. We utilize three datasets in our experiments: MovieLens-1M (Movie)[1], Book-Crossing (Book)[2], and LastFM-1K (Music)[3]. The statistics of three datasets are list in Table 1. We preprocess the data by removing inactive users and items. Users with fewer than 10 records and items consumed by fewer than 10 (50 for the Music data) users are removed.

Account Synthesis. We create synthetic accounts by merging several users' histories together. We follow [16,30] and set the distribution of the number of users in an account as $\{p_1 = 0.33, p_2 = 0.22, p_3 = 0.15, p_4 = 0.10, p_5 = 0.07, p_6 = 0.05, p_7 = 0.03, p_8 = 0.02, p_9 = 0.01, p_{10} = 0.01\}$, where p_i is the percentage of account that contains i user, and each user only belongs to one account.

Evaluation Metrics. For evaluating the performance of user identification on historical data, we use Normalized Mutual Information (NMI) on the k-means clustering results. The k-NN prediction result on new-coming sessions is measured by the average accuracy.

---

[1] https://grouplens.org/datasets/movielens/1m/.

[2] http://www2.informatik.uni-freiburg.de/~cziegler/BX/.

[3] http://ocelma.net/MusicRecommendationDataset/lastfm-1K.html.

Table 2. Results of the recommendation performances on three datasets.

Dataset	@	metric	BPR	WRMF	CoFactor	ExpoMF	APR
Movie	10	Precision	4.220%	4.070%	4.070%	3.930%	4.190%
		Recall	1.638%	1.548%	1.548%	1.403%	1.595%
		NDCG	0.0625	0.0464	0.0464	0.0517	0.0617
	20	Precision	4.495%	4.125%	4.120%	4.090%	4.225%
		Recall	3.521%	2.985%	2.979%	2.920%	3.288%
		NDCG	0.0674	0.0523	0.0522	0.0573	0.0675
Book	10	Precision	0.560%	0.560%	0.560%	0.630%	0.480%
		Recall	1.364%	1.422%	1.422%	1.193%	1.224%
		NDCG	0.0146	0.0120	0.0119	0.0120	0.0101
	20	Precision	0.490%	0.570%	0.570%	0.570%	0.500%
		Recall	2.251%	2.619%	2.621%	2.285%	2.245%
		NDCG	0.0176	0.0174	0.0174	0.0170	0.0156
Music	10	Precision	8.307%	4.265%	4.277%	4.931%	7.632%
		Recall	0.942%	0.433%	0.433%	0.520%	0.843%
		NDCG	0.1097	0.0476	0.0477	0.0528	0.0824
	20	Precision	7.071%	3.823%	3.834%	4.519%	6.705%
		Recall	1.553%	0.881%	0.884%	0.970%	1.468%
		NDCG	0.0918	0.0431	0.0432	0.0491	0.0742
Dataset	@	**metric**	**NeuMF**	**CDAE**	**CFGAN**	**SA-BPR**	**Improv.**
Movie	10	Precision	3.900%	4.130%	4.240%	**5.690%**	34.198%
		Recall	1.550%	1.543%	1.360%	**2.216%**	35.320%
		NDCG	0.0574	0.0603	0.0485	**0.0731**	17.084%
	20	Precision	3.920%	4.130%	4.060%	**5.020%**	11.680%
		Recall	2.887%	3.183%	2.542%	**3.810%**	8.203%
		NDCG	0.0622	0.0661	0.0512	**0.0737**	9.285%
Book	10	Precision	0.600%	0.480%	0.370%	**0.860%**	36.508%
		Recall	1.293%	1.029%	0.920%	**1.667%**	17.212%
		NDCG	0.0134	0.0120	0.0071	**0.0189**	29.331%
	20	Precision	0.490%	0.450%	0.370%	**0.650%**	14.035%
		Recall	2.065%	1.785%	1.671%	**2.710%**	3.406%
		NDCG	0.0166	0.0153	0.0097	**0.0234**	32.754%
Music	10	Precision	7.368%	7.048%	6.785%	**9.233%**	11.157%
		Recall	0.823%	0.773%	0.614%	**1.023%**	8.662%
		NDCG	0.0930	0.0911	0.0780	**0.1153**	5.056%
	20	Precision	6.150%	6.207%	5.688%	**7.426%**	5.016%
		Recall	1.366%	1.346%	1.202%	**1.620%**	4.294%
		NDCG	0.0783	0.0777	0.0577	**0.0955**	3.990%

Table 3. User identification results in historical and new-coming data.

Dataset	Movie		Book		Music	
	Historical	New	Historical	New	Historical	New
Metric	NMI	Accuracy	NMI	Accuracy	NMI	Accuracy
DeepWalk	0.4714	0.3751	0.8096	0.5408	0.7731	0.8889
LINE	0.5268	0.4288	0.5688	0.3550	0.6321	0.7752
metapath2vec	0.4049	0.2835	0.6128	0.5562	0.5949	0.8503
BGEM	0.4088	0.2966	0.5063	0.4922	0.4779	0.8597
SHE-UI	0.4388	0.3214	0.4701	0.2653	0.5397	0.7495
PDAE	**0.5756**	**0.4962**	**0.9062**	**0.5891**	**0.8672**	**0.8965**

To measure the recommendation performance, we employ two relevance-based metrics - Precision@K and Recall@K, and one ranking-based metric - NDCG@K.

Baseline Methods. For user identification, we compare the proposed PDAE model and PDAME model with several well-known network representation learning algorithms, which include DeepWalk [23], LINE [25], BGEM [34], and metapath2vec [3]. Besides, an embedding-based state-of-the-art algorithm for user identification, SHE-UI [16] is included. For item recommendation, we select several state-of-the-art ranking-based algorithms as baselines, which include the traditional matrix factorization-based methods: BPR [24], WRMF [15], ExpoMF [18], and CoFactor [17]; and the deep recommenders: APR [13], NeuMF [14], CDAE [31], and CFGAN [2].

5.2   Evaluation of User Identification

Implementation Details. All baseline methods except SHE-UI learn the item embedding from the complete heterogeneous information network that contains all fields listed in Table 1. SHE-UI is the only algorithm that learns the item representation solely on the sessions and item meta graph.

For each account, the k-means algorithm is applied to its session features to group sessions to N clusters, i.e., users. Without loss of generality, we set N to be the ground truth number of users of the account, which makes the evaluation results more intuitive to comprehend. For the new-coming session in an account, we get its session feature and then use k-NN to find K nearest neighbors for it. The new session is assigned to the user who has the maximum neighbor sessions.

We set the embedding size to 128. The learning rate of the Adam optimizer is 0.008 and the mini-batch size is 8192. The maximum length of the random walk sequence is set to be 10 and the context window size is 2. The neighbor size in k-NN is 10. These settings are chosen by the grid search on the validation set.

Performance Evaluation. The identification results of PDAE, PDAME and the baseline methods are shown in Table 3. It shows that PDAE outperforms all baseline methods on all metrics. We can gain several insights from Table 3.

First, The performance of BGEM is inferior to random walk-based methods (DeepWalk, LINE, metapath2vec), which indicates the preference information in the network structure is not captured well. SHE-UI achieves the worst performance because it abandoned the user information altogether.

Second, the results on Music are significantly better than those on Movie and Book datasets. The reason is obvious: Movie and Book datasets are rating data, which means there is at most one record for all user-item pairs. The rating data makes identification hard because no item repeats itself in a user's consuming logs. In contrast, a user would repeat playing his/her favorite songs in Music, which is conducive to confirming the right user.

Visualization. For a better understanding of the effect of the proposed model, we visualize the learned user embedding vectors of PDAE in the Book data via t-SNE [21], as shown in Fig. 2. The results show that users with the same profile categories are grouped together as expected. It should be noted that the color grey and the shape circle represent the unknown value.

5.3   Evaluation of Item Recommendation

Implementation Details. We build and train the proposed SA-BPR algorithm based on the identification results on historical data, where N models are built in an account. The evaluation is conducted after the recommendation is produced for the identified user on new-coming sessions. Other baseline methods assume that each account is owned by only one user so the recommender systems are trained on all sessions of an account without distinguishing users. Therefore, the baseline methods produce the same recommendation result for all users behind a shared account.

Performance Evaluation. The recommendation results are listed in Table 2, where SA-BPR outperforms all baseline methods significantly on three datasets.

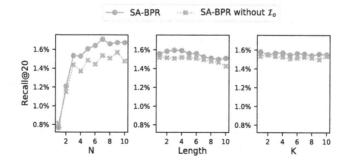

Fig. 3. Effects of parameters for item recommendation.

It indicates that identifying users does have positive effects on the user experience for shared account-aware item recommendation. Even though some identification results are wrong, SA-BPR still performs better than baseline algorithms that do not distinguish users altogether. The performance of BPR is superior to other baseline algorithms, which indicates that the personalized ranking strategy is suitable for the shared account situation.

Parameter Analysis. We explore the effects of three parameters on the recommendation performance of SA-Ranking. Limited by space, we only evaluate the Music data. The results are shown in Fig. 3. The most important insight is that the performance improves with the growth of the clustering number N. It indicates that the large clustering number N harms user identification but is conducive to personalized recommendation. The reason is intuitive: While the large N increases the chance of making mistakes in user identification, it models the user preferences in finer granularity so that each identified user (virtual personality) would have a purer preference.

The other parameters, i.e., the length of the new-coming session L and the number of neighbors in k-NN K, do not affect the performance significantly. It means that even if there are misidentified users, the recommendation result can still fulfill users' needs in that the accurate preference of the recognized virtual personality (instead of the right user) has been successfully captured.

Besides, we remove the items set $\mathcal{I}_o$ in SA-Ranking to explore the performance change. It shows that the performance degenerated, which demonstrates the effectiveness of the proposed SA-Ranking. Note that SA-Ranking without $\mathcal{I}_o$ becomes the normal BPR model but it still works on the identification result and is not like the BPR in Table 2 which does not distinguish users.

6   Conclusion

In this work, we investigate the problem of recommending items for individual users behind shared accounts. We decompose the problem into two tasks: user identification and item recommendation. For user identification, we propose the PDAE model. The core idea of PDAE is to learn the item representation and distinguish users by capturing both user preferences and demographic information. k-means and k-NN are employed to identify users and assign sessions on the historical data and new-coming data, respectively. For item recommendation, the SA-Ranking model is proposed. A novel point of SA-Ranking is that it is fault-tolerant since it provides a remedy for misidentified cases in user identification and learns parameters by the confidence of user identification. The experimental results demonstrate that the proposed models outperform all baseline methods in both user identification and item recommendation. An important insight is that even if some users are misidentified, we can still achieve satisfactory results on item recommendation because the preferences of the identified virtual personalities have been captured and modeled in finer granularity.

References

1. Aharon, M., Hillel, E., Kagian, A., Lempel, R., Makabee, H., Nissim, R.: Watch-it-next: a contextual tv recommendation system, ECML PKDD 2015, pp. 180–195 (2015)
2. Chae, D.K., Kang, J.S., Kim, S.W., Lee, J.T.: CFGAN: a generic collaborative filtering framework based on generative adversarial networks, CIKM 2018, pp. 137–146. (2018)
3. Dong, Y., Chawla, N.V., Swami, A.: Metapath2vec: scalable representation learning for heterogeneous networks, KDD 2017, pp. 135–144 (2017)
4. Dong, Y., Chawla, N.V., Tang, J., Yang, Y., Yang, Y.: User modeling on demographic attributes in big mobile social networks. ACM Trans. Inf. Syst. (TOIS) **35**(4), 35 (2017)
5. Gao, C., et al.: Alleviating matthew effect of offline reinforcement learning in interactive recommendation. In: Proceedings of the 46th International ACM SIGIR Conference on Research and Development in Information Retrieval, SIGIR 2023, pp. 238–248. Association for Computing Machinery, New York, NY, USA (2023). https://doi.org/10.1145/3539618.3591636
6. Gao, C., Lei, W., He, X., de Rijke, M., Chua, T.S.: Advances and challenges in conversational recommender systems: a survey. AI Open **2**, 100–126 (2021). https://doi.org/10.1016/j.aiopen.2021.06.002, https://www.sciencedirect.com/science/article/pii/S2666651021000164
7. Gao, C., et al.: Kuairec: a fully-observed dataset and insights for evaluating recommender systems. In: Proceedings of the 31st ACM International Conference on Information and Knowledge Management, CIKM 2022 (2022). https://doi.org/10.1145/3511808.3557220
8. Gao, C., et al.: Kuairand: an unbiased sequential recommendation dataset with randomly exposed videos. In: Proceedings of the 31st ACM International Conference on Information and Knowledge Management, CIKM 2022 (2022). https://doi.org/10.1145/3511808.3557624
9. Gao, C., et al.: Cirs: Bursting filter bubbles by counterfactual interactive recommender system. ACM Trans. Inf. Syst. (2023). https://doi.org/10.1145/3594871, just Accepted
10. Gao, C., Yuan, S., Zhang, Z., Yin, H., Shao, J.: BLOMA: explain collaborative filtering via boosted local rank-one matrix approximation. In: Li, G., Yang, J., Gama, J., Natwichai, J., Tong, Y. (eds.) DASFAA 2019. LNCS, vol. 11448, pp. 487–490. Springer, Cham (2019). https://doi.org/10.1007/978-3-030-18590-9_72
11. Guo, L., Tang, L., Chen, T., Zhu, L., Nguyen, Q.V.H., Yin, H.: Da-GCN: a domain-aware attentive graph convolution network for shared-account cross-domain sequential recommendation. In: IJCAI (2021)
12. Guo, L., Zhang, J., Chen, T., Wang, X., Yin, H.: Reinforcement learning-enhanced shared-account cross-domain sequential recommendation. IEEE Trans. Knowl. Data Eng. **35**, 7397–7411 (2022)
13. He, X., He, Z., Du, X., Chua, T.S.: Adversarial personalized ranking for recommendation, SIGIR 2018, pp. 355–364 (2018)
14. He, X., Liao, L., Zhang, H., Nie, L., Hu, X., Chua, T.S.: Neural collaborative filtering, WWW 2017, pp. 173–182. (2017)
15. Hu, Y., Koren, Y., Volinsky, C.: Collaborative filtering for implicit feedback datasets, ICDM 08, pp. 263–272 (2008)

16. Jiang, J.Y., Li, C.T., Chen, Y., Wang, W.: Identifying users behind shared accounts in online streaming services, SIGIR 2018, pp. 65–74 (2018)
17. Liang, D., Altosaar, J., Charlin, L., Blei, D.M.: Factorization meets the item embedding: regularizing matrix factorization with item co-occurrence, RecSys 2016, pp. 59–66 (2016)
18. Liang, D., Charlin, L., McInerney, J., Blei, D.M.: Modeling user exposure in recommendation, WWW 2016, pp. 951–961 (2016)
19. Ma, M., Ren, P., Lin, Y., Chen, Z., Ma, J., Rijke, M.d.: π-net: a parallel information-sharing network for shared-account cross-domain sequential recommendations, SIGIR 2019, pp. 685–694. (2019)
20. Maas, A.L., Hannun, A.Y., Ng, A.Y.: Rectifier nonlinearities improve neural network acoustic models. In: ICML Workshop 2013 (2013)
21. Maaten, L.V.D., Hinton, G.: Visualizing data using t-sne. J. Mach. Learn. Res. **9**, 2579–2605 (2008)
22. Mikolov, T., Chen, K., Corrado, G., Dean, J.: Efficient estimation of word representations in vector space. In: ICLR 2013 (2013)
23. Perozzi, B., Al-Rfou, R., Skiena, S.: Deepwalk: online learning of social representations, KDD 2014, pp. 701–710 (2014)
24. Rendle, S., Freudenthaler, C., Gantner, Z., Schmidt-Thieme, L.: BPR: bayesian personalized ranking from implicit feedback, pp. 452–461. UAI 2009 (2009)
25. Tang, J., Qu, M., Wang, M., Zhang, M., Yan, J., Mei, Q.: Line: large-scale information network embedding, WWW 2015, pp. 1067–1077 (2015)
26. Twardowski, B.: Modelling contextual information in session-aware recommender systems with neural networks, RecSys 2016, pp. 273–276 (2016)
27. Veličković, P., Cucurull, G., Casanova, A., Romero, A., Liò, P., Bengio, Y.: Graph attention networks. In: ICLR 2018 (2018)
28. Verstrepen, K., Goethals, B.: Top-n recommendation for shared accounts, RecSys 2015, pp. 59–66 (2015)
29. Wang, Z., He, L.: User identification for enhancing IP-tv recommendation. Knowl.-Based Syst. **98**, 68–75 (2016)
30. White, R.W., Hassan, A., Singla, A., Horvitz, E.: From devices to people: attribution of search activity in multi-user settings, WWW 2014, pp. 431–442 (2014)
31. Wu, Y., DuBois, C., Zheng, A.X., Ester, M.: Collaborative denoising auto-encoders for top-n recommender systems, WSDM 2016, pp. 153–162 (2016)
32. Yang, S., Sarkhel, S., Mitra, S., Swaminathan, V.: Personalized video recommendations for shared accounts, ISM 2017, pp. 256–259 (2017)
33. Yang, Y., Hu, Q., He, L., Ni, M., Wang, Z.: Adaptive temporal model for IPTV recommendation, WAIM 2015, pp. 260–271 (2015)
34. Yin, H., Zou, L., Nguyen, Q.V.H., Huang, Z., Zhou, X.: Joint event-partner recommendation in event-based social networks, ICDE 2018, pp. 929–940 (2018)
35. Zhang, A., Fawaz, N., Ioannidis, S., Montanari, A.: Guess who rated this movie: identifying users through subspace clustering, UAI 2012, pp. 944–953 (2012)
36. Zhao, Y., Cao, J., Tan, Y.: Passenger prediction in shared accounts for flight service recommendation, APSCC 2016, pp. 159–172 (2016)

Prediction and Characterization of Social Media Communication Effects of Emergencies with Multimodal Information

Yuyang Tian[1], Shituo Ma[2], Qingyuan He[3], and Ran Wang[1(✉)]

[1] School of Journalism and Information Communication,
Huazhong University of Science and Technology, Wuhan, Hubei, China
`rex_wang@hust.edu.cn`
[2] School of Computer Science and Technology,
Huazhong University of Science and Technology, Wuhan, Hubei, China
[3] School of Chemistry and Chemical Engineering,
Huazhong University of Science and Technology, Wuhan, Hubei, China

Abstract. This paper investigates the communication effects on social media, and their influencing factors. A emergency events dataset spanning the period from 2019 to 2023 is constructed, comprising a large volume of textual and image data obtained through web crawling. The communication effects of emergency social media posts during various emergent events are analyzed using a comprehensive paradigm, with the breadth and depth of dissemination measured by the sum of likes and comments, as well as the number of reposts. LightGBM is employed as the classifier, and multidimensional features incorporating visual and textual dimensions are constructed. Experimental results highlight the significant impact of the image modality on dissemination effects, particularly emphasizing the importance of features such as HSV and image content categories. Additionally, the number of followers of the original poster is identified as a crucial factor influencing dissemination effects. The experimental results show that the research method based on feature engineering and machine learning can effectively predict the propagation effect of microblog, and the LightGBM algorithm performs best. The study further found that in the comparison of modal effects, the graphical multimodal has better performance.

Keywords: Communication Effect · Multimodal Information · LightGBM · Emergency

1 Introduction

In recent years, due to the ability of multimodal content, such as images and videos, a majority of social media posts have become increasingly visualized. Surveys have shown that over 51.6% of social media posts include images, and the

F. Wu et al. (Eds.): SMP 2023, CCIS 1945, pp. 77–89, 2024.
https://doi.org/10.1007/978-981-99-7596-9_6

reposting volume of social media posts with images exceeds that of purely textual posts by a factor of 10. Therefore, predicting the propagation effects of multi-modal breaking event social media posts can help management departments detect potential issues in a timely manner and enhance the foresight of decision-making.

However, existing research lacks attention to multi-modal features such as images, and there has been insufficient utilization of natural language processing and computer vision techniques in studying the dissemination effects of social media, especially in the context of breaking event content. Additionally, the existing measurement metrics pose challenges and are not suitable for studying the dissemination effects primarily driven by multi-modal content.

This study aims to focus on social media posts related to breaking events that contain multi-modal information and their dissemination effects. It will employ machine learning and deep learning techniques to achieve the following research objectives:

1. Firstly, a prediction model for dissemination effects based on multi-modal features was constructed. By utilizing multi-level feature classification algorithms from machine learning methods, model can investigate the evolutionary patterns and dissemination mechanisms of various feature factors to achieve the most accurate prediction of dissemination effects.
2. Secondly, the influencing factors of dissemination effects in the context of breaking events were deeply analysed. By comprehensively considering the multi-level features of text and images, model will systematically analyze the impact of these factors on the dissemination process and scientifically elucidate the intrinsic patterns of information dissemination in social media.

These works will enable timely identification of potential issues and enhance the proactive decision-making of management departments.

2   Related Works

2.1   Evaluation of Communication Effects

In the field of communication studies, media influence can be defined as the ability of media to influence the attitude system of the audience and other related actors [1]. Communication effect refers to the degree of changes in cognition, attitudes, and behaviors of the target audience caused by the communicator's communication activities.

Research based on media effects has decomposed the elements of media communication, which include the communication sender, the audience, the content, the channels, and interactive indicators. Changes at the cognitive level can be measured through indicators such as website traffic and the number of unique visitors [2]. The behavioral level of change is the most important aspect in the study of communication effects, and commonly used online measurement metrics. Depending on different communication objectives and specific circumstances, other indicators such as conversion rates and transaction rates may also be included [3]. In studies based on influence, influence is defined as a comprehensive reflection of breadth, depth, and validity. Breadth refers to the audience reach and can be measured by indicators such as total views. Depth [4] can be measured by indicators such as likes and comments. Validity indicators [5,6] cover the duration of user discussions and reposts.

Existing research has measured communication effects from different dimensions, but in terms of specific measurement metrics, the platform's built-in interactive features (such as reposts, likes, and comments) are representative methods for evaluating the dissemination effects of social media information, with only a few studies employing manual coding.

2.2   Factors Affect Communication Effect

Current research on the factors influencing communication effects mainly focuses on three categories: sender characteristics, audience characteristics, and content characteristics. Sender characteristics include popularity, authority, and activity. Audience characteristics include demographic features, activity, and interest. Content characteristics are mainly divided into three categories: content type, content quality, and content interaction. Emotional features include emotional intensity, whether the content expresses positive emotions, and whether it contains extreme emotions. Relevance can be measured by the similarity between text topics and user interests, as well as the similarity between user emotions and social media emotions. Freshness can be determined by the time of content publication and the number of comments, likes, and shares in the user's previous interaction. These features can help us better understand the popularity of content.

However, these influencing factors have limitations. First, many indicators may be inaccessible. For example, for private or unverified accounts, we may not have access to their number of followers, mentions, citations, or direct messages. Second, the specific features of multi-modal data such as text, images, or videos are not simultaneously considered. For example, features such as image clarity, originality, color, and composition are not adequately addressed. Therefore, this paper will integrate multiple features of text and images to achieve more accurate prediction results and supplement existing research.

2.3   Multimodal Information in Communication

Multi-modality refers to the inclusion of various content formats, including but not limited to text, images, sound, videos, animations, and charts, encompassing several sensory modalities such as visual, auditory, and tactile [7]. On one hand, due to the character limit in texts, users often include multiple content formats to assist in conveying information. On the other hand, compared to traditional single-modal media, multi-modal forms such as images and videos have a stronger visual impact, increasing the expressiveness of content and enabling it to gain broader dissemination and attention. Currently, multi-modal content has become mainstream, with over 51.6% of social media posts including images, and it has been found that social media posts with images have over 10 times the repost volume compared to purely textual posts [8]. This significant difference in dissemination effects highlights the crucial role of images in the information dissemination process. Based on the analysis above, it is necessary to comprehensively utilize visual content, such as images, to assist in detecting communication effects.

The current research landscape on the utilization of multimodal information in social media applications encompasses two primary categories of research questions. The first category focuses on employing image feature recognition techniques to identify rumors and false news [9]. Within this domain, researches have unveiled the correlation between news images and events, the usage of cartoon images, and the freshness of images, all of which impact the credibility of news content [10]. The second category of notable research inquiries pertains to predicting the propagation effectiveness of content [11]. Such studies predominantly examine dependent variables including interactive data such as the number of reposts, likes, and comments for video or image content on social media [12]. Additionally, a subset of studies explores audience sentiment and reactions. The range of factors under scrutiny in these studies is diverse, spanning content-related characteristics, attributes of content creators (e.g., their number of followers), aesthetic qualities of images (e.g., clarity, originality, exposure) [13], visual traits of human facial portraits [14], and features related to poses and movements [15, 16].

3   Dataset Construction

A series of previous research cases have demonstrated that mining emergency events data from social media can accurately achieve risk warning. Therefore, the dataset will be crawled from texts on social media, a Chinese social media platform. Texts of 135 official accounts from January 1, 2009 to April 8, 2023 were selected, processed and annotated. Events include many types of subordinate events like public health, social security, natural disaster and crowd accident. The richness of events is high, including 42 types of emergencies.

Based on the research paradigms of microblog dissemination effects both domestically and internationally, this study adopts the "influence" paradigm, representing the dissemination effects of individual microblogs in terms of dissemination breadth and dissemination depth. This paradigm utilizes the interactive data of microblogs, where dissemination breadth is measured by the sum of the number of likes and comments, while dissemination depth is represented by the number of reposts. The overall representation of dissemination effects is the sum of dissemination breadth and dissemination depth. This choice is based on the understanding that likes, reposts, and comments are three actions that individuals can take to interact with a post, and each of these actions alone cannot fully reflect people's reactions to the microblog post.

Figure 1 shows some statistical feature of dataset.

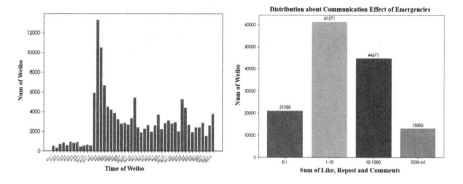

Fig. 1. Statistical results of similarity

As can be seen from the Fig. 1, January to February 2020, were the dates with the highest number of breaking news tweets published, with more than 10,000 per month, while all months in 2019 had significantly fewer breaking news tweets published than other years and months, with less than 1,000 tweets. In addition, since the Sina Weibo platform allows different numbers of images to be inserted when posting tweets, it is necessary to explore the distribution of the number of images included in tweets. The distribution of the number of images shows a long-tail trend, with the vast majority of microblogs having only 1 image and very few microblogs with more than 9 images as the long-tail, where the maximum number of images can reach 18. The distribution of communication effects of different microblogs is shown, showing the number of microblogs whose communication effects, i.e., the sum of the number of likes and comments, fall in three intervals. It can be seen that the most number of microblogs have a spreading effect between 1 and 10 interactions; less than one tenth of the microblogs reach more than 1000 interactions, and it is worth noting that there are still one seventh of the microblogs without any spreading effect.

4   Communication Effect Prediction Model

4.1   Feature Selection

To discuss the impact of multi-modal features on the dissemination effect data of social media, this study classifies these features into three categories: sender features, content features, and temporal features. For each category, representative and important sub-features are selected. Based on content features, multi-modal features are further divided into text and image categories to explore the effects of different types of features on interaction data. The feature selection is based on recent domestic and international research as well as the analysis of social media interaction data, providing interpretability and influence in the context of social media interaction data. Features are listed in Table 1:

Table 1. Feature engineering encoding results

Main Feature	No.	Feature	Sub-feature	Range
Sender	1	num of fans		5410–150973200
	2	num of posts		4385–184032
	3	VIP-level		0–7
Texual	4	num of '@'		0–11
	5	num of '#'		1–10
	6	is original		0/1
	7	classification	Main type	0–3
			Subtype	0–48
	8	Length		8–6004
	9	Emotion Strength		0–1
	10	TF-IDF		1–10
Visual	11	num of images		0–18
	12	HSV	H	0–180
	13		S	0–255
	14		V	0–255
	18	image type	class	128d word vec
	19		object	128d word vec
	20	human faces	persons	0–5
	21		gender	0/1
	22		age	0–94
	23		expression	0–100
	24		emotion	0–7
	25		appearance	0–100
Time	26	year		2019–2022
	27	month		1–12
	28	day		1–31
	29	hour		0–24

On the Weibo platform, content features are crucial factors in determining the values of likes, reposts, and comments. In this study, content features are divided into text and image categories.

Regarding text content features, they can be categorized into pragmatic features and statistical features based on the semantic abstraction level. Three pragmatic features are selected: originality, text classification, and text sentiment, along with four statistical features: the number of "@" mentions, the number of "#" hashtags, text length, and TF-IDF text word vectors. In pragmatic features, "originality" indicates the uniqueness and originality of social media content. Text classification features are extracted using the pre-trained BERT model to investigate the performance of social media dissemination effects under different topics. Additionally, text sentiment features provide quick insights into the emotional tendencies of social media posts, helping to evaluate their influence and promptly address public opinion risks.

In terms of image content features, they can be divided into high-level features and low-level features based on their interpretability. High-level features have higher interpretability and align more closely with human cognition. They include facial features and image types that express the meaning of images intuitively. This study selects three high-level features: HSV color features, facial features, and image types. A worth mentioning feature is HSV feature. HSV represents a color space and includes three components: hue, saturation, and value. Here, HSV is a one-dimensional vector of length 25, including 16 hue values, 8 saturation values, and 1 value. Facial features include the number of individuals, gender, age, expressions, emotions, and attractiveness in the image.

Regarding sender features, we select the number of followers, number of posts, and VIP level of social media posters, as shown in Table 1. These features reflect the influence, engagement, and status of social media users. The number of followers and VIP level of social media users are important indicators of their influence and reputation, while the number of posts reflects their activity level on the social media platform. Previous research has shown a positive correlation between the number of followers, number of posts, VIP level, and social media interaction data.

In terms of time, we selected year, month, day and hour as four features to describe time. Through the analysis of time, we can discuss infections during special time or special events.

4.2   Model Architecture

Based on the features above, the multi-modal communication effect prediction model that integrates text and image features can be divided into four modules: the input module, feature representation module, feature fusion module, and output module, as shown in Fig. 2.

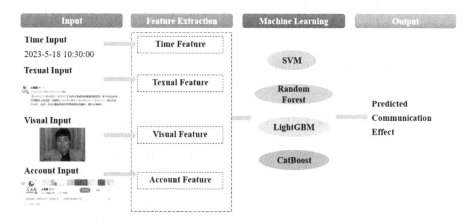

Fig. 2. Model architecture for communication effect prediction.

From the Fig. 2 we can see that there are four modules:

1. Firstly, the input module extracts the relevant information from the raw social media data, including the original data corresponding to each feature and the preprocessing applied to them. The preprocessed text and image data are then used as input for the model.
2. Next, in the feature representation and fusion module, different methods are employed to achieve vectorized representations of the preprocessed data. This includes the representation of sender features, multi-modal content features, and temporal features.
3. In the machine learning model training module, four classifiers, namely Support Vector Machine (SVM), Random Forest (RF), LightGBM, and CatBoost, are utilized to train the fused feature vectors.
4. Finally, in the output module, the prediction results of the multi-modal dissemination effect detection model are computed.

Overall, this model integrates text and image features, allowing for a comprehensive analysis of the multi-modal nature of social media posts and their impact on dissemination effects.

To solve the problem of heterogeneity of image features, this study takes the following measures:

1. Regarding the heterogeneity caused by different feature extraction models, only the relevant computed dimensions are retained while removing redundant information such as thresholds and confidences.
2. To address the heterogeneity in the number of feature vectors resulting from different image characteristics, which may vary due to the number of faces or objects in the images, this study selects the highest quality face features and the most reliable object scene recognition results. This ensures that each image corresponds to a single feature vector.
3. Considering the heterogeneity caused by the varying number of images in each social media post, padding strategy is employed. Padding refers to the process of adding additional values to the edges of data in order to ensure consistent input data shape in computer vision. Specifically, this study pads the image features of each social media post with zero values based on the maximum number of images in a social media post, thus ensuring consistent shapes for all social media image features.

Four multi-classification models were tested in this study: SVM, Random Forest, CatBoost, and LightGBM. SVM is a classification algorithm based on maximum margin, suitable for high-dimensional data and non-linear classification problems. Random Forest is an ensemble learning algorithm based on decision trees, capable of effectively handling high-dimensional data and non-linear relationships. CatBoost and LightGBM are improved versions of gradient boosting decision trees, capable of handling high-dimensional data and large-scale datasets. They have advantages in training efficiency and prediction accuracy. CatBoost uses a special symmetric tree structure that handles class imbalance issues in classification problems better. LightGBM employs histogram-based acceleration and gradient-based sparse feature combination techniques, greatly improving training efficiency and prediction accuracy.

Finally, the performance of the multi-modal dissemination effect detection model is evaluated based on precision, recall, and F1-measure metrics. The comprehensive performance is compared to assess their predictive effectiveness.

5   Experiment and Evaluation

5.1   Classification Performance

To verify and compare the performance of image unimodal, text unimodal and graphical multimodal in the task of predicting the propagation effect of unexpected events by go-betweens. The experimental results obtained using different modal information compared with different model combinations with the corresponding parameters kept constant are shown in Table.2.

Table 2. Classification Performance

Model	P	R	F1
SVM + Texual	0.909	0.757	0.789
Random Forest + Texual	0.918	0.918	0.913
CatBoost + Texual	0.909	0.909	0.908
LightGBM+Texual	0.917	0.917	0.913
SVM + Visual	0.826	0.768	0.796
Random Forest + Visual	0.884	0.901	0.892
CatBoost + Visual	0.899	0.917	0.908
LightGBM+Visual	0.919	0.919	0.914
Random Forest + Multimodal	0.914	0.908	0.911
LightGBM+Multimodal	**0.921**	**0.921**	**0.916**

According to the experimental results in Table.2, LightGBM is the model with the best results. It is shown that the precision, recall and F1 of image modality are better than text modality, while the performance of graphical multimodality is between text and image modality, indicating that image modality has better performance in predicting the propagation effect of tweets for breaking events. The performance of the graphical multimodal achieves better results compared to the single modality, proving that multimodal information has better results in the propagation effect prediction task.

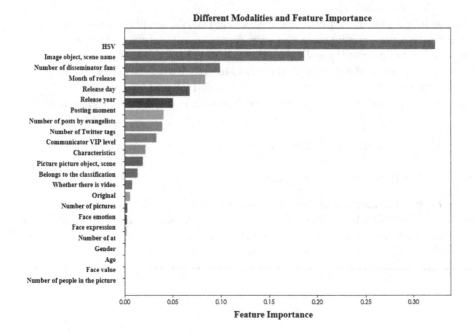

Fig. 3. Feature Importance in LightGBM.

5.2   Feature Importance

After automatic feature engineering of the LightGBM model, the model gives feature importance scores for different features. The importance of different features in the model is plotted in the Fig. 3:

From Fig. 3, some interesting findings can be seen:

Firstly, the predictive performance of the image modality is better than the text modality and the multimodal combination of text and image. This result indicates that images have a stronger influence on the dissemination of information during the spread of emergency events compared to text, and the combination of text and image does not provide complementary effects. When images contain rich visual information and can directly convey the message, users tend to skip redundant textual descriptions and focus directly on the images. This can be attributed to the higher cognitive demands of language comprehension compared to the visual system. Visual communication benefits from the natural representation of images, as images have the ability to reproduce real-world phenomena, which is lacking in linguistic symbols. Studies have shown that in crisis situations, clear visual presentations facilitate the public's understanding of crisis response decisions.

In terms of feature importance analysis, HSV features have the most significant impact on the dissemination of emergency event microblogs, followed by the secondary image classification feature, which represents the content category of images. This result demonstrates the crucial role of color and content information conveyed by images in promoting the dissemination of emergency events, which is consistent with the principles of semiotics.

The experimental results indicate that the number of followers/fans of the disseminators is the third most influential factor on the dissemination effects. This can be understood based on the initial exposure range and the subsequent cascade effects of content dissemination on social media platforms. During the cold-start phase of emergency event microblogs, the content is initially propagated through social networks, which means that disseminators with a larger follower base have an advantage in terms of initial exposure. Moreover, microblogs with high initial exposure are more likely to receive continuous promotion and exposure through content recommendation mechanisms and trending lists. Therefore, the number of followers/fans plays a foundational role in the dissemination effects of microblogs.

Similarly, besides the number of followers/fans, the number of posts and VIP status of the disseminators also have a certain impact on the dissemination effects. The number of posts and VIP status reflect the activity level of disseminators on the microblog platform. Hence, the regular activity level of disseminators also influences the final dissemination effects of emergency event microblogs. However, in this study, the facial features exhibit very weak importance in terms of their impact on dissemination effects.

6 Conclusion and Future Works

To explore the important role of multimodal information in the communication process, a multimodal information communication model was constructed by extracting two categories of features from the original data divided into two categories of communicators and content as the influencing factors of communication effect. The data of official media microblog users on Sina Weibo website are crawled with more than 140,000 items and more than 130,000 images are processed, and multiple machine learning models are used for comparison training. A social media communication effect prediction method applicable to multimodal content is proposed, and the importance of different features on communication effect prediction is comprehensively evaluated to provide insights for public opinion guidance and management.

Although the model has made preliminary analytical work on multimodal information, the following issues remain to be further explored:

1. Feature alignment processing enhancement: on the microblogging platform, each microblog contains a different number of pictures, so the picture features and microblogging features are in a many-to-one relationship. The accuracy and completeness of feature alignment is improved when performing the fusion of picture features and the features of the microblogs to which they belong.
2. Feature fusion: deep learning methods such as attention mechanism are used to form a more abstract feature fusion.

3. Feature interpretation needs to be strengthened: Although the machine learning algorithm used in this study can produce more accurate prediction results, the interpretation of the model needs to be further explored.

Acknowledgement. This research is founded by National Social Science Fund of China (19CXW032).

References

1. Liyong, Z., Danni, Z., Chun, Z.: Research on media influence evaluation index system. Journal. Univ. **01**, 121–126 (2010)
2. Artzi, Y., Pantel, P,. Gamon, M.: Predicting Responses to Microblog Posts[C/OL] Conference of the North American Chapter of the Association for Computational Linguistics: Human Language Technologies, pp. 602–606. Association for Computational Linguistics, Montréal, Canada (2012)
3. Ye, W., Jia, F., Lun, Z.: An examination of the communication effect of short video personification in mainstream media-an analysis of the visual content based on the column "Anchors Say Link". J. Xi'an Jiaotong Univ. (Soc. Sci. Edi.) **41**(2), 131–139 (2021)
4. Suh, B., Hong, L., Pirolli, P., et al.: Want to be retweeted? Large scale analytics on factors impacting retweet in twitter network. In: IEEE Second International Conference on Social Computing, pp. 177–184 (2010)
5. Liu, X., Zheng, W.: Analysis of the effect of corporate social media marketing communication-Taking microblog diffusion network as an example. Journal. Commun. Res. **22**(02), 89–102+128 (2015)
6. Chunhui, H., Jiayu, M.: Quantitative indicators of media influence. Journal. Pract. **10**, 15–17 (2006)
7. Ning, H. Lin, Y., Cheng, W.: An empirical analysis of factors influencing the communication effect of major public health emergencies: the example of health and health Jitterbug government number. Mod. Commun. (J. Commun. Univ. China), **43**(01), 147–151 (2021)
8. Qiang, C., Yating, P.: Research on the factors influencing the information dissemination effect of governmental Jitterbug numbers in major public health emergencies: an empirical analysis based on 25 official Jitterbug numbers of provincial health care commissions. J. Guangxi Normal Univ. (Philos. Soc. Sci. Ed.) **58**(04), 72–85 (2022)
9. Chen, S., Dou, Y., Zhang, Q.: Research on the influencing factors of microblog users' retweeting behavior based on rational behavior theory. J. Intell. **36**(11), 147–152+160 (2017)
10. Shiyu, D., Jiayin, Q.: Evaluation of microblog topic influence index based on principal component analysis. J. Intelligence **33**(05), 129–135 (2014)
11. Kaur, J., Jain, D.: A study of gender-based differences in social media usage. Int. J. Inf. Educ. Technol. **5**(5), 445–448 (2015)
12. Wei, L., Min, H., Lihong, W., et al.: Research on microblog retweet prediction based on user behavior features. J. Comput. Sci. **39**(10), 1992–2006 (2016)
13. Li, C.-Q.: Microblog retweet prediction based on LDA topic features. J. Intell. **34**(09), 158–162 (2015)
14. Liao, L., Huang, T.: Exploring the influence of sources, information content and emotional characteristics on microblog retweets. Mod. Intell. **40**(9), 42–52 (2020)

15. Linna, X.I., Yongxiang, D.O.U.: Research on the influencing factors of microblog users' retweeting behavior based on the theory of planned behavior. Data Anal. Knowl. Discov. **3**(02), 13–20 (2019)
16. Jun, X.: Multimodal discourse analysis: a theoretical model and its methodological implications for cross-cultural communication research in new media. J. Wuhan Univ. (Human. Ed.) **70**(06), 126–134 (2017)

Leverage Heterogeneous Graph Neural Networks for Short-Text Conceptualization

Xiaoye Ouyang[1], Yashen Wang[1,2(✉)], Qiang Li[3], Zhuoya Ju[3], Chenyu Liu[1], and Yi Zhang[1,4]

[1] National Engineering Laboratory for Risk Perception and Prevention (RPP), China Academy of Electronics and Information Technology, Beijing 100041, China
yswang.arthur@gmail.com
[2] Key Laboratory of Cognition and Intelligence Technology (CIT), Artificial Intelligence Institute of CETC, Beijing 100144, China
[3] Scientific and Technological Innovation Center of ARI, Beijing 100012, China
[4] CETC Academy of Electronics and Information Technology Group Co., Ltd., Beijing 100144, China

Abstract. The conceptualization of short-texts is playing an increasingly important role in text comprehension, social media processing and other applications. Generally, this problem could be modeled as a heterogeneous semantic network connecting words (in short-text) and corresponding concepts (in knowledge base). As a crucial step in short-text conceptualization task, learning interactive relations among words and concepts has been explored through numerous methods. One intuitive method is to place them in a graph based neural network with a more complex structure to capture inter-concept/word relationships. Hence, this paper presents a heterogeneous graph-based neural network (HGNN) for short-text conceptualization, which includes semantic nodes with various granularity levels, mainly consisting of basic- semantic nodes (e.g., words) and supernodes (e.g., concepts). The proposed model could provide a flexible and natural modeling tool to model such complex relationships and capture more expressive and discriminative concepts, by leveraging mutually reinforcing strategy on heterogeneous correlations. On the other word, it is a beneficial attempt to introduce different types of semantic nodes into graph based neural networks for short-text conceptualization task, and we conduct comprehensive qualitative analysis to investigate its benefits.

Keywords: Short-Text Conceptualization · Heterogeneous Graph Neural Networks · Social Media Analysis

1 Introduction

Short-text conceptualization plays an increasingly critical role in text understanding, social media analysis and other applications [7,21,22,26–28,30]. It is

F. Wu et al. (Eds.): SMP 2023, CCIS 1945, pp. 90–103, 2024.
https://doi.org/10.1007/978-981-99-7596-9_7

an interesting task to project a piece of short-text to a set of pre-defined concepts with different granularities.

Recent work on understanding short-texts has placed greater emphasis on using signals from lexical knowledge bases to assist in conceptualization of short texts [7,9,27,31], and achieved great success. Many *probabilistic* and *co-ranking* based algorithms have been proposed and gradually become a state-of-the-art model [9,20,21,30]. However, in recent years, there has been *little* taste in using depth graph neural networks in this task, which has been applied for many real-world applications. In layman's terms, given a short-text (e.g., "microsoft unveils office for apple's ipad") as input, we aim at mapping each word to one or several candidate concepts which have been defined in extra lexical knowledgebase (e.g., the widely-used Probase [31]), and construct the heterogeneous semantic network (as shown in the middle part of Fig. 1, wherein ellipses indicate concepts defined in Probase, and rectangles indicate words extracted from the current short-text). Apparently, in such a semantic network, relationships are complex and go beyond conventional pairwise associations. E.g., many words may co-refer a same concept (e.g., word "apple" and "microsoft" are all subordinate to concept COMPANY); a word could be a co-subordination of more than two concepts in this semantic network (e.g., word "apple" belongs to concept FRUIT and concept COMPANY). A more intuitive approach for this structure, is to model the interactions between words and concepts using *graph structures*. However, finding an effective graph structure for conceptualizing short-texts is challenging as well as the key point. Objectively speaking, Graph Neural Networks (GNN) provide a flexible and natural modeling tool to model such complex relationships, Showcasing the potential to address the aforementioned issues. The apparent existence of this complex relationship naturally have stimulated the problem of graph learning [6,12,13,15,24,32].

Therefore, this paper would like to investigate how effective is the application of GNNs towards short-text concept conceptualization task, and tries to leverage a graph neural network inspired by [24] respect to heterogeneous semantic graph for solving short-text conceptualization. Intuitively, we introduce multiple types of semantic vertices in this kind of heterogeneous semantic graph mentioned above, including *word vertices* and *concept vertices*, which enriches the interactive relationships among different semantic signals embedded in the current short-text. Besides, each word is connected to its corresponding concepts, and edge weight can be easily derived from Probase [9]. During the process of massage passing and aggregation through heterogeneous graphs and graph attention manoeuvre, word vertices and concept vertices will be iteratively updated, similar to the idea behind [9]. Finally, we utilize a sequence labeling procedure to draw concept vertices into the final concept set for representing the given short-text. Correlation among concepts could be easily captured under our modeling criteria.

Our contributions can be concluded as follows: (i) It is a beneficial taste to construct a HGNN paradigm for releasing short-text's concept representation, by deeply modeling the mutual relations between words and concepts. This model enables the heterogeneous signals (i.e., words and concepts) to sufficiently interplay, by leveraging mutually reinforcing strategy for modeling heterogeneous and bidirectional correlations among words and concepts. (ii) The proposed framework is very flexible in terms of extension and can be easily adjusted from other kinds of semantic units, apart from concepts emphasized here, in the future. (iii) Experiments such as qualitative analysis etc., have proven the effectiveness of our model, and moreover our model can surpass all existing competitors in real-world datasets/tasks without pre-training language models[1].

2   Task Definition

2.1   Concept

Following previous work [21,25], this paper also defines a "concept" as a set or class of "entities" or "things" within a domain, such that words belonging to same (or similar) classes get similar representations. Note that, Probase [31] is used in this paper as the lexical knowledgebase as well as the vocabulary, which has been widely used in previous research about short-text conceptualization.

2.2   Short-Text Conceptualization

Given a short-text $s = \{w_1, \cdots, w_m\}$ with a number of m words, we can obtain a concept set C_{w_i} for each word w_i by utilizing instance conceptualization API provided by [31] and eliminating the concept whose API score is less than 0.001 for computational efficiency. E.g., given word "microsoft", the instance conceptualization algorithm will return the corresponding concept distribution and different confidence level, as follows: $C_{microsoft}$={COMPANY, VENDOR, CLIENT, FIRM, ORGANIZATION, CORPORATION, BRAND, $\cdots$ }. Then, we could obtain the set of candidate concepts for the current short-text s by $V_c = C_{w_1} \bigcup \cdots \bigcup C_{w_m}$. For simplicity, $V_c = \{c_1, \cdots, c_n\}$ with size of n, wherein, c_j indicates the candidate concept driven from Probase for representing the entire short-text s. With efforts above, we can formulate short-text conceptualization as a sequence labeling task, as follows: Our aim is to reason a sequence of labels $\{\mathbb{I}_1, \cdots, \mathbb{I}_n\}$ for candidate concepts, wherein $\mathbb{I}_j$ beyond the threshold presupposed means the j-th candidate concept c_j should be retained in the final optimal concept set for represent the given short-text s.

3   Data Modeling

Inspired by the task definitions (Sect. 2), mathematically, short-text conceptualization task forms a naturally heterogeneous graph G, which consists two types of vertices: (i) word's vertex and (ii) concept's vertex (similar to the definition proposed in [24].

---

[1] Since the proposed model does not contradict the method of using pre-training models, we think that our model can be further improved by using pre-training architecture to initialize vertex representations, and we reserve it for the future.

Formally, we can represents this kind of heterogeneous graph as $G = \{V, E\}$, wherein: (i) notation V indicates the vertex set and $V = V_w \bigcup V_c$ with $V_w = \{w_1, \cdots, w_m\}$ (respect to given short-text s) and $V_c = \{c_1, \cdots, c_n\}$ (corresponds to the n candidate concepts derived from Probase (details in last Sect. 2)) standing for word's vertex set and candidate concept's vertex set respectively; notation E indicates the edge set representing edges between vertices and $E = \{E[i,j] | i \in \{1, \cdot, m\}, j \in \{1, \cdots, n\}$ with $E[i,j]$ as element of matrix-formed E to indicate the undirected and weighted correlation between word w_i and concept c_j. Note that, $E[i][j] \neq 0$ indicates the i-th word w_i belongs to j-th candidate concept c_j. That is, word w_i is mapped to concept c_j in Probase.

4 Methodology

Figure 1 overviews the proposed architecture, which includes three modules: (i) graph initialization for vertices and edges (Sect. 4.1), (ii) heterogeneous attention processing (Sect. 4.2), and (iii) final concept selection (Sect. 4.3).

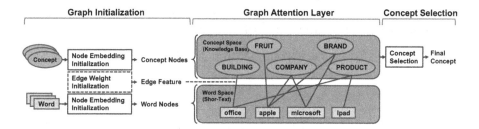

Fig. 1. The sketch of the proposed model.

4.1 Graph Initialization

This section summaries the initialization procedure of vertex vector and edge weights in G, respectively.

Node Embedding Initialization. Let $\mathbf{W} \in \mathbb{R}^{m \times d_w}$ and $\mathbf{C} \in \mathbb{R}^{n \times d_c}$ describe the input embedding matrix of word's vertices and concept's vertices independently, with d_w indicating the dimension of the word representation vector and d_c indicating the dimension of concept representation vector. Specifically, notation $\mathbf{w}_i$ indicates the vector representation for word w_i, and is initialized by conventional Word2Vec model [17][2]. Then, we leverage $\mathcal{P}(w_i|c_j)$ score for generating concept's vector representation $\mathbf{c}_j$ for each concept c_j. Wherein, $\mathcal{P}(w_i|c_j)$ denotes the probability of word w_i given concept c_j, which can be derived from

---

[2] Note that, other pre-trained word embedding methods could also be adapted here.

Probase conveniently [31]. Hence, for simplicity, for current concept c_j, its initialized vector representation can be generated as the following simple way: $\mathbf{c}_j = \sum_{E[k][j] \neq 0} \mathcal{P}(w_k | c_j) \cdot \mathbf{w}_k$. With efforts above, the inputs of the first layer in our GNN can uniformly be formed as: $\mathbf{h}_i^{(0)} \leftarrow \mathbf{w}_i$ is v_i is a word, while $\mathbf{h}_i^{(0)} \leftarrow \mathbf{c}_i$ is v_i is a concept.

Edge Weight Initialization. For initializing edge's weight $E[i][j] \in E$ embedded in G, to further include information about the importance of mutual relationships between word and concept vertices, this work tries to seek help from $\mathcal{P}(c_j | w_i)$ score in the edge weights, which has depicted the probability of candidate concept c_j given word w_i. $\mathcal{P}(c_j | w_i)$ can be also intuitively generated by the instance conceptualization algorithm [20].

4.2 Graph Attention Processing

After graph initialization step, a typical Graph Attention (GAT) Network [23] is leveraged here for synergistically updating the vector representations our semantic vertices embedded in $\mathbf{W}$ and $\mathbf{C}$. We then denote $\mathbf{h}_l \in \mathbb{R}^{d_h}$ ($l \in \{1, \cdots, |V|\}$) as the hidden states for each vertex $v_l \in V$, and the message passing and aggregation mechanism of l-th GAT layer can be designed as follows:

$$z_{i,j}^{(l)} = \sigma_1(\mathbf{M}_2^{(l)}[\mathbf{M}_1^{(l)}\mathbf{h}_i \,\|\, \mathbf{M}_1^{(l)}\mathbf{h}_j^{(l)}]) \tag{1}$$

$$a_{i,j}^{(l)} = \frac{\exp(z_{i,j}^{(l)})}{\sum_{j' \in \mathcal{N}_i} \exp(z_{i,j'}^{(l)})} \tag{2}$$

$$\mathbf{h}_i^{(l+1)} = \sigma_2\Big(\sum_{k \in \mathcal{N}_i} a_{i,k}^{(l)} \mathbf{W}_3^{(l)} \mathbf{h}_k^{(l)}\Big) \tag{3}$$

Wherein, function $\sigma_1(\cdot)$ and $\sigma_2(\cdot)$ act as the non-linear activation function here [23] [16]. Operation $\|$ represents concatenation and $\mathcal{N}_i$ indicates the neighbor vertex collection of current vertex v_i. $a_{i,j}^{(l)}$ is the attention weight between $\mathbf{h}_i^{(l)}$ and $\mathbf{h}_j^{(l)}$ respect to l-th layer, which adaptively assigns different importance to different neighboring vertices in $\mathcal{N}_i$.

Based on Eq. 3, an interactive update procedure [24] is introduced here for mutually reinforcing word vectors and concept vectors, shown in Fig. 2. For simplicity, we denote the hidden state matrix as $\mathbf{H} \in \mathbb{R}^{(m+n) \times d_h}$, which consists of two sub-matrices $\mathbf{H}_w \in \mathbb{R}^{m \times d_h}$ and $\mathbf{H}_c \in \mathbb{R}^{n \times d_h}$ representing hidden states respect to word's vertices and concept vertices individually. Among the total T iterations, each iteration includes a *word-to-concept* update procedure and a *concept-to-word* update procedure. For the t-th iteration ($t \in T$), the process can be defined as follows:

Word→Concept Updates: The word-to-concept update process in the l-th layer's t-th iteration is formulated as follows:

$$\mathbf{H}_c^{(l,t+1)} \leftarrow \text{GAT}(\mathbf{H}_w^{(l,t)}, \mathbf{H}_c^{(l,t)}, \mathbf{H}_c^{(l,t)}) \tag{4}$$

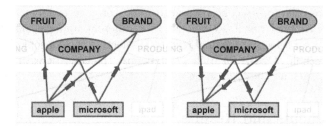

Fig. 2. Sketch of interactive update procedure respect to word vertices (yellow rectangles) and concept vertices (red ellipses), including word-to-concept update process (left side) and concept-to-word update process (right side): E.g., (i) for concept COMPANY, word "apple" and word "microsoft" are used to aggregate information, as shown in the left side; (ii) word "microsoft" is updated by the new vector representation of its corresponding concept COMPANY and BRAND, as shown in the right side. (Color figure online)

Wherein, $\mathtt{GAT}(\mathbf{H}_w^{(l,t)}, \mathbf{H}_c^{(l,t)}, \mathbf{H}_c^{(l,t)})$ denotes that $\mathbf{H}_w^{(l,t)}$ is used as the attention *query* and $\mathbf{H}_c^{(l,t)}$ is used as the *key* and *value*.

Concept→Word Updates: The concept-to-word update process in the l-th layer's t-th iteration is formulated as follows:

$$\mathbf{H}_w^{(l,t+1)} \leftarrow \mathtt{GAT}(\mathbf{H}_c^{(l,t)}, \mathbf{H}_w^{(l,t+1)}, \mathbf{H}_w^{(l,t+1)}) \tag{5}$$

The main intuition behind the above strategy is that there is a *mutually reinforcing* relationship between concepts and words, which can be reflected in heterogeneous graphs. Hence, similar to the co-ranking based short-text conceptualization model [9], the proposed model could also simultaneously generate the keywords for the current short-text s. Regarding candidate concept vertices ($c_j \in V_c$), the one with more important words tends to be selected into the final concept set (details in following Sect. 4.3). In a word, therefore, increasing the ranking of words and their concepts relies on each other in a mutually reinforcing manner, thereby utilizing the additional information implicit in the heterogeneous graph composed of words and concepts. The aforementioned mutual reinforcing procedure typically converges when difference between the vector representations of concept vertices computed at two successive iterations, falls below a certain threshold. Note that, other choice for judging converging can be used here, e.g., when the effective loss does not decrease within three or four consecutive iterations, a stop can be executed.

4.3 Optimal Concept Selection

Section 2 has formulated short-text conceptualization as a sequence labeling task. Hence, we utilize a sequence labeling procedure based on the final hidden state $\mathbf{H}$, to draw concept vertices into the final concept set for representing the given short-text, wherein cross-entropy loss is used as the training objective. And finally we choose the top-n' concepts and top-m' words according to descending ranking-scores as final results. Obviously, for different applications, there are other strategies for selecting the final result: (i) selecting the concept of exceeding the preset threshold; Or (ii) for each word, choose the concept with the highest ranking score, such as [30], and so on[3].

5   Experiments and Results

5.1   Alternative Algorithms

We compare the our Heterogeneous Graph Neural Network based short-text conceptualization model (denoted as **HGNN**), with the following short-text conceptualization algorithms: (i) the state-of-the-art probabilistic algorithms [20, 30], which are considered as strong baselines; (ii) the state-of-the-art co-ranking based algorithms; (iii) currently proposed graph neural network learning based algorithms, which could be transferred into short-text conceptualization task easily and be most comparable to our work; as well as (iv) basic-baselines such as bag-of-words based algorithms weighted by TF-IDF scores and LDA's topic distributions, acting as basic baselines.

BOW: [19] represents given short-text as bag-of-words with the TF-IDF scores, which is viewed as the most intuitive and simplest methodology.

LDA: [1] represents given short-text as its inferred topic distribution, and the dimensions of the short-text vector of is number of topics as presupposed. This algorithm regards "topic" as "concept".

IJCAI$_{11}$: As an early job for short-text conceptualization, [20] proposes a probabilistic framework, which implements a concise co-clustering of concepts (derived from Probase) and words by identifying the disjoint cliques, and then derives the most likely concepts using Bayesian inference. Note that, all the following alternative algorithms leverage concepts defined by knowledgebase Probase for this task, as well as ours.

IJCAI$_{15}$: As a representative work in domain of short-text understanding, [30] tries to conceptualizes words using a random-walk based iterative algorithm, by taking verbs and adjectives into consideration.

Co-Rank$_{AD}$: [33] proposes a co-ranking framework by simply coupling two random walks, which separately rank different type of semantic units, such as vertices under PageRank's paradigm [2] focussed by our conceptualization task.

Co-Rank$_{HITS}$: [11] proposes Hyperlink Induced Topic Search, which could used in this task due to its ability of coping with heterogeneous information by confirming ranking could be reinforced through interactions.

Co-Rank$_{IP}$: Based on a novel co-ranking framework, [9] simultaneously ranks the concepts and words in an iterative procedure, which can be regarded as the strong baseline in the line of research about co-ranking strategy.

MPHGNN: [15] proposes Meta-Path-based HGNN (MPHGNN) model, which has been adopted for our conceptualization task by using a multi-head

---

[3] We ignore the further discussion here since it is not the focus of this paper.

attention mechanism to aggregate signals on neighbors of different kinds of vertices (i.e., words and concepts emphasized here).

HAEGNN: [13] proposes Higher-order Attribute-Enhancing GNN (HAE GNN) model, which is applied here for leveraging the self-attention mechanism for exploring content-based word and concept vertices' interactions in short-text conceptualization.

5.2 Datasets

For text clustering task to evaluate the similarity among vectors represented by concepts generated by the comparative, three widely-used datasets are introduced here, including **NewsTitle, Twitter** and **TREC**:

NewsTitle: We extract news's titles from a news collection consisting of 3.62 million articles searched from Reuters and New York Time. By manually labeling, the news articles are split into several categories as follows: *company, religion, science, traffic, politician,* and *sport.* The standard dataset is split into 70%/10%/20% for training, validating and testing. Then we randomly select 5,000 news titles in each category.

Twitter: We utilize the official tweet collections used in TREC Microblog Task 2011/2012 [18] to draw tweets (in form of short-text natively) and construct this dataset. By manually labeling, the dataset contains 12,378 tweets which are in following categories: *food, sport, entertainment,* and *device/IT company.* The standard dataset is split into 70%/10%/20% for training, validating and testing. We remove the URLs and stop-words. The average length of the tweets is 12.95 words. Because of noise and sparsity, this dataset is more challenging.

TREC: this collection is respect to question clustering task on TREC [14], which is widely used as benchmark for conceptualization task. The entire dataset of 5,952 sentences with the following categories: *person, entity, abbreviation, description, location* and *numeric.* Similarly, we also split the standard dataset into 70%/10%/20% for training, validation and test.

Moreover, the datasets used for word similarity task, will be described latter in the corresponding sections.

5.3 Experiment Settings

Due to space limitations, we will briefly introduce the experimental setups here.

We limit the vocabulary to Probase (similar to previous research). The numbers of word' vertices and concept's vertices, m and n, is naturally derived from the corresponding short-texts (as well as concepts activated in Probase) from different datasets. The dimensionality of concept vector is set as $d_c = 128$, the dimensionality of concept vector is set as $d_w = 128$, and the dimensionality of hidden state vector is $d_h = 64$. The threshold for label $\mathbb{II}_j$ annotated for concept c_j in sequence of candidate concepts, is cautiously set of 0.6 in this work. The total number of interactive update procedure T proposed in Sect. 4.2 is set as 3, 5 and 3 respect to different datasets about news, tweets and questions, respectively. Similar the total number of GNN's layers L used in Sect. 4.2 is set as

10, 20 and 15 respect to the above datasets, respectively, which have released the optimal results. During procedure of training the proposed model, Adam optimizer is applied with a learning rate $5e - 4$. For decoding, we select top-10 concepts for TREC dataset (i.e., $n' = 10$), and top-15 for NewsTitle/Twitter dataset (i.e., $n' = 15$), according to the average length of their text.

For all algorithms, the Porter algorithm extracts the stem of each short-text, and the stopwords are removed from the InQuery tool's stopword-list. The dimension of a vector in **BOW** model is 50,000; For **LDA** model, we set the topic number to cluster number or twice, and report the better of the two. To improve efficiency, for **Co-Rank$_{AD}$**, **Co-Rank$_{HITS}$**, **Co-Rank$_{IP}$**, **IJCAI$_{11}$**, **IJCAI$_{15}$**, **HAEGNN**, **MPHGNN** and ours, we follows [29] and select ~5,000 concepts as features in text clustering task, which is like the number of concept clusters in Probase [9,25,30].

5.4   Experiments on Short-Text Clustering

Due to the lack of real labels for news title data, tweet data, and question data, we designed a short-text clustering task to evaluate the effectiveness of conceptualization. In summary, we first generate the concepts (pre-defined in Probase or in form of topics) of each short-text based on different algorithms, then use these concepts as features to generate a short-text vector, and finally run a spherical K-means clustering [3] to evaluate each algorithm. We report the average metric Normalized Mutual Information (NMI) scores [5] of 20 random trials for comparison (as shown in Table 1).

Table 1. Performance of short-text clustering task.

Model	NewsTitle	Twitter	TREC
BOW [19]	0.781	0.435	0.863
LDA [1]	0.683	0.297	0.760
IJCAI$_{11}$ [20]	0.807	0.436	0.875
IJCAI$_{15}$ [30]	0.822	0.440	0.882
Co-Rank$_{HITS}$ [11]	0.749	0.427	0.859
Co-Rank$_{AD}$ [33]	0.806	0.439	0.874
Co-Rank$_{IP}$ [9]	0.879^{ψ}	0.478^{ψ}	0.942^{ψ}
MPHGNN [15]	0.815	0.435	0.877
HAEGNN [13]	0.845	0.443	0.889
HGNN (Ours)	$\mathbf{0.904}^{\gamma\psi}$	$\mathbf{0.492}^{\gamma\psi}$	$\mathbf{0.950}^{\gamma\psi}$

The results from Table 1, show the proposed short-text conceptualization model based on our heterogeneous graph neural network (**HGNN**) improves the baseline models in most cases: (i) **HGNN** exceeds the best baseline model **IJCAI$_{15}$** by 10.34%, **IJCAI$_{11}$** by 12.39%, **Co-Rank$_{AD}$** by 12.53%,

Co-Rank$_{IP}$ by 3.19% and **HAEGNN** by 7.34% in **NewsTitle** dataset; and (ii) **HGNN** exceeds **IJCAI$_{15}$** by 12.73%, **IJCAI$_{11}$** by 13.76%, **Co-Rank$_{AD}$** by 12.98%, **Co-Rank$_{IP}$** by 3.77% and **HAEGNN** by 11.96% in **Twitter** dataset. Note that, the statistical t-test is employed here, and the superscript γ and ψ respectively denote statistically significant improvements over **Co-Rank$_{IP}$** and **HAEGNN** ($p < 0.05$). This indicates that the proposed **HGNN** can provide a flexible and natural modeling instrument to simulate complex relationships and capture more expressive and discriminative concepts by utilizing mutual reinforcement strategies regarding heterogeneous correlations. This phenomenon can be explained as the model effectively utilizes the beneficial interactions between words and concepts, allowing signals to fully interact while being influenced by text noise and sparsity (especially compared with **IJCAI$_{15}$** and **LDA**).

The results adequately demonstrate the effectiveness of mutual-ranking between concepts and words, by the comparison among probabilistic algorithms (e.g., **IJCAI$_{11}$** and **IJCAI$_{15}$** etc.,), co-ranking based algorithms (e.g., **Co-Rank$_{AD}$**, **Co-Rank$_{HITS}$** and **Co-Rank$_{IP}$** etc.,) and the proposed **HGNN**. Besides, our advantage compared with co-ranking frameworks (**Co-Rank$_{AD}$**, **Co-Rank$_{HITS}$** and **Co-Rank$_{IP}$**) indicates that graph attention processing as well as signals passing and aggregation (details in Sect. 4.2) is valuable, and the affinity matrix could better reflect the strength of correlation. We also examine the number of iterations. Experimental results shows that compared to other collaborative ranking algorithms, **HGNN** uses fewer iterations to achieve convergence.

5.5 Qualitative Analysis

This section aims at investigating the influence of the vertex degree of words on the conceptualization result. Since in the proposed model, *higher* degree words can aggregate signals from more various kinds of concepts, which means they can benefit *more* from the iterative procedure (discussed in Sect. 4.2).

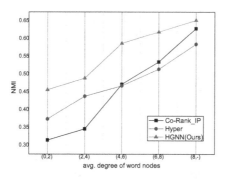

Fig. 3. Effects of average degree of word's vertices.

We evaluate the performance of **HGNN**, **HAEGNN** and **Co-Rank$_{IP}$** on NewsTitle dataset, with 5 kinds of average degree of word vertices (i.e., (0, 2), (2, 4), (4, 6), (6, 8) and more than 8). Experimental results shown in Fig. 3 conclude that: (i) Mainly targeting all comparative models, NMI scores increase as the average degree of word vertices in the context increases; (ii) **Co-Rank$_{IP}$** has been particularly affected with degree diversity of word vertices, and our **HGNN** performs much better in case with a higher value of average word vertex degree.

5.6   Experiments on Word Similarity

This paper introduces experiments on word similarity following [10], wherein the proposed models and comparative models are applied to measure the similarity between words presented in short-texts. A common dataset for this task, dataset **WordSim-353** [4] consists of pairs of words without context. However, as we can see in the "apple" example, the meanings of homophones and polysemy may vary depending on the context. [8] proposes a new dataset to measure the similarity between words in context, where the dataset provides two words in sentences and ten non-experts annotate their similarity scores. After reading each pair of excerpts, the annotator evaluates the similarity of highlighted words on a scale of 0 to 10. In this experiment, we conceptualize the highlighted words using comparison models, and then measure the cosine similarity between the concept distributions of two words.

Table 2. Performance of word similarity task.

Model	Pearson-Correlation
BOW	0.357
LDA	0.289
IJCAI$_{11}$ [20]	0.329
IJCAI$_{15}$ [30]	0.588
Co-Rank$_{HITS}$ [11]	0.536
Co-Rank$_{AD}$ [33]	0.577
Co-Rank$_{IP}$ [9]	0.617^{ψ}
MLP [15]	0.583
HAEGNN [13]	0.604
HGNN (Ours)	$\mathbf{0.654}^{\gamma\psi}$

For evaluation purposes, we have calculated the Pearson's correlation between model-based judgment and human judgment. We calculate the cosine similarity between the BOWs of two target words to measure their similarity. Table 2 compares the results of different models and indicates that our method outperforms the baseline results. Note that, the results prove that, modeling complex relationships and enabling the signals (i.e., words and concepts) to fully interplay, improves word similarity judgements over naive conceptualization without context, especially compared with the state-of-the-art probabilistic based baselines (e.g., **IJCAI$_{11}$** and **IJCAI$_{15}$**).

6   Conclusion

This paper proposes a novel Heterogeneous Graph Neural Network (HGNN) based model to address the problem of short-text conceptualization, which could provide a flexible and natural modeling tool to understand complex relationships and capture more expressive and discriminative concepts, by leveraging mutually reinforcing strategy for modeling heterogeneous correlations among words and concepts. Experiments on real-world datasets suggest that the proposed model is effective.

Acknowledgements. We thank anonymous reviewers for valuable comments. This work is funded by: (i) the National Natural Science Foundation of China (No. 62106243, U19B2026, U22B2601).

References

1. Blei, D.M., Ng, A.Y., Jordan, M.I.: Latent dirichlet allocation. J. Mach. Learn. Res. **3**, 993–1022 (2003)
2. Brin, S., Page, L.: The anatomy of a large-scale hypertextual web search engine. In: International Conference on World Wide Web, pp. 107–117 (1998)
3. Dhillon, I.S., Modha, D.S.: Concept decompositions for large sparse text data using clustering. Mach. Learn. **42**(1), 143–175 (2001)
4. Finkelstein, L., et al.: Placing search in context: the concept revisited. ACM Trans. Inf. Syst. **20**, 116–131 (2002)
5. Fred, A.L.N., Jain, A.K.: Robust data clustering. In: IEEE Computer Society Conference on Computer Vision and Pattern Recognition, pp. 128–133 (2003)
6. Gao, C., Wang, X., He, X., Li, Y.: Graph neural networks for recommender system. In: Proceedings of the Fifteenth ACM International Conference on Web Search and Data Mining (2022)
7. Hua, W., Wang, Z., Wang, H., Zheng, K., Zhou, X.: Short text understanding through lexical-semantic analysis. In: IEEE International Conference on Data Engineering, pp. 495–506 (2015)
8. Huang, E.H., Socher, R., Manning, C.D., Ng, A.Y.: Improving word representations via global context and multiple word prototypes. In: ACL (2012)
9. Huang, H., Wang, Y., Feng, C., Liu, Z., Zhou, Q.: Leveraging conceptualization for short-text embedding. IEEE Trans. Knowl. Data Eng. **30** (2018)
10. Kim, D., Wang, H., Oh, A.: Context-dependent conceptualization. In: International Joint Conference on Artificial Intelligence, pp. 2654–2661 (2013)
11. Kleinberg, J.M.: Authoritative sources in a hyperlinked environment. J. ACM **46**(5), 604–632 (1999)
12. Lamb, L., Garcez, A., Gori, M., Prates, M., Avelar, P., Vardi, M.: Graph neural networks meet neural-symbolic computing: a survey and perspective (2020)

13. Li, J., et al.: Higher-order attribute-enhancing heterogeneous graph neural networks. IEEE Trans. Knowl. Data Eng. **35**, 560–574 (2022)
14. Li, X., Roth, D.: Learning question classifiers. In: 19th International Conference on Computational Linguistics, pp. 1–7 (2002)
15. Liang, X., Ma, Y., Cheng, G., Fan, C., Yang, Y., Liu, Z.: Meta-path-based heterogeneous graph neural networks in academic network. Int. J. Mach. Learn. Cybern. **13**, 1553–1569 (2021)
16. Maas, A.L.: Rectifier nonlinearities improve neural network acoustic models (2013)
17. Mikolov, T., Chen, K., Corrado, G., Dean, J.: Efficient estimation of word representations in vector space. Computer Science (2013)
18. Ounis, I., Macdonald, C., Lin, J.: Overview of the trec-2011 microblog track (2011)
19. Salton, G., Mcgill, M.J.: Introduction to modern information retrieval. McGraw-Hill (1983)
20. Song, Y., Wang, H., Wang, Z., Li, H., Chen, W.: Short text conceptualization using a probabilistic knowledgebase. In: Proceedings of the Twenty-Second International Joint Conference on Artificial Intelligence - Volume Volume Three, pp. 2330–2336 (2011)
21. Song, Y., Wang, S., Wang, H.: Open domain short text conceptualization: a generative + descriptive modeling approach. In: International Conference on Artificial Intelligence, pp. 3820–3826 (2015)
22. Sun, Y., Yang, Y., Yang, D.: Informed graph convolution networks for multilingual short text understanding. In: International Conference on Knowledge-Based Intelligent Information & Engineering Systems (2022)
23. Velickovic, P., Cucurull, G., Casanova, A., Romero, A., Liò, P., Bengio, Y.: Graph attention networks. ArXiv abs/1710.10903 (2018)
24. Wang, D., Liu, P., Zheng, Y., Qiu, X., Huang, X.: Heterogeneous graph neural networks for extractive document summarization. In: Annual Meeting of the Association for Computational Linguistics (2020). https://api.semanticscholar.org/CorpusID:216552978
25. Wang, F., Wang, Z., Li, Z., Wen, J.R.: Concept-based short text classification and ranking. In: The ACM International Conference, pp. 1069–1078 (2014)
26. Wang, Y., Huang, H., Feng, C.: Query expansion based on a feedback concept model for microblog retrieval. In: International Conference on World Wide Web, pp. 559–568 (2017)
27. Wang, Y., Huang, H., Feng, C., Zhou, Q., Gu, J., Gao, X.: Cse: conceptual sentence embeddings based on attention model. In: 54th Annual Meeting of the Association for Computational Linguistics, pp. 505–515 (2016)
28. Wang, Y., Liu, Y., Zhang, H., Xie, H.: Leveraging lexical semantic information for learning concept-based multiple embedding representations for knowledge graph completion. In: APWeb/WAIM (2019)
29. Wang, Y., Zhang, H.: Harp: a novel hierarchical attention model for relation prediction. ACM Trans. Knowl. Discov. Data **15**, 17:1–17:22 (2021)
30. Wang, Z., Zhao, K., Wang, H., Meng, X., Wen, J.R.: Query understanding through knowledge-based conceptualization. In: International Conference on Artificial Intelligence, pp. 3264–3270 (2015)
31. Wu, W., Li, H., Wang, H., Zhu, K.Q.: Probase: a probabilistic taxonomy for text understanding. In: Proceedings of the 2012 ACM SIGMOD International Conference on Management of Data, pp. 481–492 (2012)

32. Yadati, N., Nimishakavi, M., Yadav, P., Nitin, V., Louis, A., Talukdar, P.P.: Hyper-gcn: a new method of training graph convolutional networks on hypergraphs. In: NeurIPS 2019 (2018)
33. Zhou, D., Orshanskiy, S.A., Zha, H., Giles, C.L.: Co-ranking authors and documents in a heterogeneous network. In: 7th IEEE International Conference on Data Mining, pp. 739–744 (2007)

Short-Text Conceptualization Based on Hyper-Graph Learning and Multiple Prior Knowledge

Li Li[1], Yashen Wang[2,3]($\boxtimes$), Xiaolei Guo[2], Liu Yuan[2], Bin Li[4],
and Shengxin Xu[2]

[1] School of Computer, Beijing Institute of Technology, Beijing 100081, China
[2] National Engineering Laboratory for Risk Perception and Prevention (RPP),
China Academy of Electronics and Information Technology, Beijing 100041, China
yswang.arthur@gmail.com
[3] Key Laboratory of Cognition and Intelligence Technology (CIT),
Artificial Intelligence Institute of CETC, Beijing 100144, China
[4] Beijing Finefurther Digital Technology Co., Ltd., Beijing 100083, China

Abstract. Short-text conceptualization is a notable task and popular issue in current social network analysis and natural language processing. This line of work usually views the data as a heterogeneous semantic network connecting terms (in short-text) and corresponding concepts (in prior knowledge base), with complex relationships (e.g., term-correlation, concept-correlation and subordination, etc.,). Therefore, this paper introduces hyper-graph learning strategy for solving this problem, because of its ability for modeling complex relationships. Overall, this paper proposes a novel short-text conceptualization model based on hyper-graph convolutional network. Especially, this model is capable to make the signals (i.e., terms and concepts) to be sufficiently interacted, by leveraging three prior knowledge for modeling heterogeneous correlations among terms and concepts, including: subordination prior knowledge, concept correlation prior knowledge and term correlation prior knowledge. The experimental results demonstrate that the proposed work achieves higher accuracy in short-text conceptualization task when compared with the current state-of-the-art algorithms.

Keywords: Short-Text Conceptualization · Hyper-Graph Learning · Prior Knowledge · Hyper-Graph Convolutional Network

1 Introduction

Mapping short-texts to a number of pre-defined concepts has gained many successful applications [24,25]. Hence, the problem of short-text conceptualization is important, and has attracted increasing attention.

F. Wu et al. (Eds.): SMP 2023, CCIS 1945, pp. 104–117, 2024.
https://doi.org/10.1007/978-981-99-7596-9_8

Especially, Fig. 1 sketches an example of the semantic network[1] sof short-text conceptualization problem, respect to the given short-text "microsoft unveils office for apple's ipad", which summaries a map from each term (bottom of Fig. 1) to candidate concepts (top of Fig. 1) in prior knowledge base (Probase emphasized in this paper), with multiple *complex* relation-types (blue, orange and red lines in Fig. 1). Apparently, this kind of semantic network is heterogeneous [18,19,27], which consists of three sub-networks [9]: (i) the *concept-correlation network* G_C connecting concepts, indicating correlation relationships; (ii) the *term-correlation network* G_T connecting terms, indicating co-occurrence relations, and (iii) the *subordination network* G_{TC} which tightly couples the aforementioned two previous networks together, denoting whether a term belongs to a concepts.

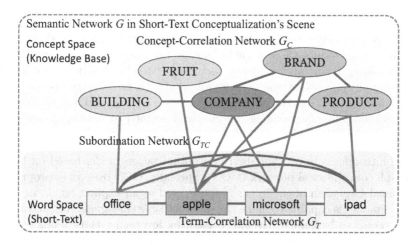

Fig. 1. An example of semantic network for short-text conceptualization problem. (Color figure online)

Hyper-graph has provided a flexible and natural modeling tool to understand such multiple and complex relationships [28,32]. Intuitively, the apparent existence of this complex relationship (in Fig. 1) naturally stimulates the problem of hyper-graph learning [3,8,16,32,36]. Therefore, this paper investigates and discusses a novel short-text conceptualization model by introducing successful hyper-graph based Graph Convolutional Network (GCN). Especially, nowadays a famous learning paradigm is semi supervised learning based on hyper-graphs, which aims to assign labels to initially unlabeled vertices in the hyper-graph [32]. Many techniques use explicit *Laplacian regularisation* in the objective [30,34]. As discussed in [5,22,32], although *explicit* Laplacian regularization assumes

---

[1] Due to space limitations, we only provide noun entities in Fig. 1. Furthermore, parsing short-text into terms or multi-term expressions may not be easy, and we overlook this point because it is not the focus of this article.

similarity between the hyper-vertices in each hyper-edge, *implicit* regularization of GCNs has avoided this limitation and can be applied to a wider range of problems [15] Hence, following the backbone of previous work [32], the proposed model tries to utilize GCN over an appropriately defined Laplacian of the hyper-graph as an *explicit* regulariser.

Moreover, previous conceptualization work on short-texts generally assumes that all observed terms (and concepts) were conditionally *independent*, and simply considered the possibility of multiplying the conditional probabilities of each term as a concept. In fact, they do a good job in mining concepts related to individual terms or publishing overly general concepts, but *fail* in mining the overall concept set of the entire short-text. In summary, in previous work, there exists an independent assumption that *overlooks* the interactions between terms and concepts (i.e., the blue and red lines in Fig. 1), and neglects the beneficial reactions of concepts to terms, which may reflect the global concept. To overcome these issues, we must design a mechanism that allows signals (i.e., terms and concepts) to fully interact and derive solid conceptualization for short-texts. Therefore, we introduce three prior knowledge for modeling heterogeneous correlations among terms and concepts, including: (i) subordination prior knowledge (in G_{TC}), concept-correlation prior knowledge (in G_C) and term-correlation prior knowledge (in G_T). The proposed mechanism is novel and flexible to evaluate terms and their corresponding concepts concurrently on heterogeneous scene.

In summary, this work contributes to the following aspects:

(i) We introduce a novel short-text conceptualization model based on hyper-graph convolutional network (GCN). Specially, GCN over an appropriately defined Laplacian of the hyper-graph is used as an explicit regulariser.

(ii) This model is capable to make the different types of signals (i.e., terms and concepts) to be sufficiently interplayed, by leveraging three kinds of prior knowledge for modeling heterogeneous correlations among terms and concepts, including: subordination prior knowledge, concept correlation prior knowledge and term correlation prior knowledge. Hence, We break the independent assumption held in previous works [20,29], and demonstrate the beneficial interactions among concepts and terms.

(iii) Experiments on real datasets have shown that this method is more effective, efficient, and flexible than other state-of-the-art methods.

2 Preliminary

Following previous definitions [23,26], this paper delimits a 'concept' as a set or class of 'entities' or 'things' within a domain, such that words belonging to similar classes get similar representations. Especially, Probase [31] is used in our study as extra knowledge base which provides high-quality pre-defined concepts.

Terms in short-text (in term space) and their corresponding candidate concepts (in concept space) form a typical heterogeneous semantic network, and we formally denote the heterogeneous semantic network as $G = (V, E)$, and $n_V = |V|$ indicates the number of all the vertices. V is the set of vertices in G and $V = V_C \cup V_T$, wherein V_C is the set of candidate concepts with size of $n_C = |V_C|$, and V_T is the set of terms with size of $n_T = |V_T|$. E indicates the set of edges and $E = E_{CC} \cup E_{TC} \cup E_{TT}$, wherein E_{CC} represents the set of edges indicating correlation ties among concepts, E_{TT} represents the set of edges among terms established by their co-occurrence relations, and E_{TC} represents the set of edges reflecting the subordination relations among terms and concepts. Overall, following [9], G is composed of three sub-networks: (i) the concept-correlation network $G_C = (V_C, E_C)$ respect to concepts, (ii) the term-correlation network $G_T = (V_T, E_T)$ respect to terms, and (iii) the bipartite subordination network $G_{TC} = (V_{TC}, E_{TC})$ that couples concepts (in G_C) and terms (in G_T) together. The affinity matrix $\mathbf{M} \in \mathbb{R}^{n_V \times n_V}$ of G is defined as follows:

$$\mathbf{M} = \begin{bmatrix} \mathbf{M_{CC}} & \mathbf{M_{CT}} \\ \mathbf{M_{TC}} & \mathbf{M_{TT}} \end{bmatrix} \tag{1}$$

wherein sub-matrix M_{CC} models the correlation relationship among concepts (in G_C), and sub-matrix M_{TT} models the co-occurrence relationship among terms. The other two sub-matrices, i.e., M_{CT} and M_{TC}, represent bipartite subordination (in G_{TC}), measuring the likelihood of a given term being assigned certain concepts, and vice versa.

3   Graph Convolutional Network

Graph Convolutional Network (GCN) [11] generally uses simple linear functions of graph Laplacian to define convolutions, and have been proven to be effective for semi-supervised learning of attribute graphs, such as classification [28,32].

In this paper, Sect. 2 has represented heterogeneous semantic network $G = (V, E)$ (consisting of terms and concepts) with affinity matrix $\mathbf{M} \in \mathbb{R}^{n_V \times n_V}$. Besides, there exists another important matrix respect to the given G, i.e., embedded matrix $\mathbf{V} \in \mathbb{R}^{n_V \times d}$, which has d-dimensional numerical vector embedded representations for each vertex (term embedded in short-text or candidate concept) $v \in V$ and each dimensionality indicates and depicts different kinds of graph signals (discussed latter). We also use notation $\mathbf{D}$ to indicate a diagonal matrix, $\mathbf{D} = \mathrm{diag}(\mathrm{D}[1][1], \cdots, \mathrm{D}[n_V][n_V])$, with elements $\mathrm{D}[i][i] = \sum_{j=1, j \neq i}^{n_V} \mathrm{M}[j][i]$ respect to each vertex $v_i \in V$ (with condition of $\mathrm{M}[j][i] \in \mathbf{M}$). Then, we can denote λ_{n_V} as the largest eigenvalue of the symmetrically-normalised graph Laplacian:

$$\mathbf{L} = \mathbf{I} - \mathbf{D}^{-\frac{1}{2}} \mathbf{M} \mathbf{D}^{-\frac{1}{2}} \tag{2}$$

Then, we can use $\hat{\mathbf{L}} = \frac{2\mathbf{L}}{\lambda_{n_V}} - \mathbf{I}$ to indicate the scaled graph Laplacian [32]. The discussions about graph convolution theorem [11,14,32] have shown that, the convolution vector $\mathbf{C} \in \mathbb{R}^{n_V}$ respect to a numerical graph signal vector $\mathbf{s} \in \mathbb{R}^{n_V}$ is approximately denoted as follows: $\mathbf{C} \approx (\lambda_1 + \lambda_2 \cdot \hat{\mathbf{L}}) \cdot \mathbf{s}$. Wherein λ_1 and λ_2 indicate weights to be trained.

Following the proofs in [32], the graph convolution for d dimensionalities (respect to d different graph signals) contained in the embedded matrix $\mathbf{V} \in \mathbb{R}^{n_V \times k}$ with trainable weights $\Theta \in \mathbb{R}^{k \times |h_2|}$ with $|h_2|$ hidden states is $\overline{\mathbf{M}}\mathbf{V}\Theta$ [11], wherein

$$\overline{\mathbf{M}} = \widetilde{\mathbf{D}}^{-\frac{1}{2}} \widetilde{\mathbf{M}} \widetilde{\mathbf{D}}^{-\frac{1}{2}} \tag{3}$$

$$\widetilde{\mathbf{M}} = \mathbf{M} + \mathbf{I} \tag{4}$$

$$\widetilde{\mathbf{D}}[i][i] = \sum_{j=1}^{n_V} \widetilde{\mathbf{M}}[i][j] \tag{5}$$

Generally, to simplify, take GCN with one hidden layers as example, Graph Convolution Operation takes the following simple form:

$$\mathbf{Z} = \texttt{SoftMax}[\overline{\mathbf{M}}\texttt{ReLU}(\overline{\mathbf{M}}\mathbf{V}\Theta_1)\Theta_2] \tag{6}$$

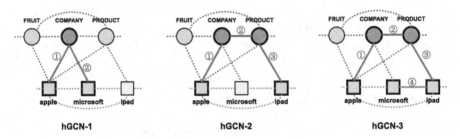

Fig. 2. The proposed hyper-Graph Convolutional Network based model for short-text conceptualization. (Color figure online)

wherein, $\Theta_1 \in \mathbb{R}^{k \times |h_1|}$ is an input-to-hidden weight matrix for a hidden layer with $|h_1|$ hidden states, and $\Theta_2 \in \mathbb{R}^{|h_2| \times |h_1|}$ is a hidden-to-output weight matrix. Weights Θ_1 and Θ_2 are trained using gradient descent.

4   The Proposed Model

This paper proposes a novel short-text conceptualization model based on hypergraph convolutional network (GCN). Especially, a novel hyper-graphs driven GCN training method [32], is introduced and verified here for our short-text conceptualization task. Methodologically, this model approximates each hyper-edge of a hyper-graph through a set of paired edges connecting hyper-edge vertices, and treats the learning problem as an approximate hyper-graph learning problem. The key element of GCN is the Laplacian graph of a given graph. Therefore, in order to develop a GCN based hyper-graph SSL method (details in Sect. 4.3), we firstly need to construct a Laplacian for the hyper-graph (details in Sect. 4.2). Finally, we describe three variants of the proposed models with different prior knowledge (details in Sect. 4.4).

4.1 Task Definition

Generally, short-text conceptualization attempts at inferring not only the optimal concept(s) for each term embedded in the given sentence (i.e., $v_t \in V_T$), but also for the whole sentence. We then could cautiously translate the short-text conceptualization problem into semi-supervised hyper-vertex classification task on an undirected hyper-graph $G = (V, E)$ with $n_V = |V_T| + |V_C|$ (simultaneously, $n_V = |V| = |V_T \cup V_C|$), $n_E = |E_{TT} \cup E_{TC} \cup E_{CC}|$. We denotes a small set $V_T^* \subset V_T$ as labelled hyper-vertices. That is, the hyper-vertices in V_T^* are all labelled by concept information. Given each edge $e \in E$, V_e indicates the vertex set containing all the vertices connecting by edge e. For each hyper-vertices $v \in V_T$, its d-dimensional feature vector is $\mathbf{v} \in \mathbb{R}^d$ and $\mathbf{V} \in \mathbb{R}^{n_V \times d}$. The task is to reason the labels (i.e., optimal concepts) of all the unlabelled hyper-vertices, that is, all the term hyper-vertices in the set $V_T \backslash V_T^*$ (i.e., $v_t \in V_T \backslash V_T^*$). Note that, the problem investigated here is a multi-classification task, wherein a term hyper-vertex could be labeled with several concepts.

4.2 Hyper-Graph Laplacian Construction

[33] has proven that hyper-vertices within the same hyper-edge are similar, so they may share the same label. Inspired by [32], this work uses hyper-graph Laplacian as an implicit regularizer to achieve the goal of hyper-vertices with similar representations within the same hyper-edge.

Therefore, in order to develop a GCN based semi-supervised learning method for hyper-graphs (consisting of terms and concepts), we firstly need to define and construct the Laplacian of the hyper-graph, which is regarded as the very circal point of GCN of a given graph $G = (V, E)$. Following [4,16], given a numerical signal vector $\mathbf{s} \in \mathbb{R}^{n_V}$ (discussed in Sect. 3), the non-linear graph Laplacian $\mathbb{L}(\mathbf{s})$ can be computed as follows:

Given a hyper-edge $e \in E$, "$v_i, v_j \in e$" indicates that hyper-vertex $v_i \in V$ and hyper-vertex $v_j \in V$ belong to hyper-edge e, i.e., both of them (the term occurred in given short-text or the corresponding candidate concept) appear in the hyper-edge e. Hence, for each hyper-edge $e \in E$, [32] denotes the two farthest hyper-vertices on the same hyper-edge e as $(v_{i(e)}, v_{j(e)}) := \arg\max_{v_i, v_j \in e} \|s[i] - s[j]\|_l$ (wherein $l = 2$), which randomly breaks ties following [4,16]. Note that, $\max_{v_i, v_j \in e} \|s[i] - s[j]\|_l$ will be *small* only if vectors corresponding to the hyper-vertices in hyper-edge e are semantic *closer* to each other.

A weighted graph $G^\star$ on the vertex set V is constructed by adding edges $\{(v_{i(e)}, v_{j(e)}) | e \in E\}$ with learnable weight $w[(v_{i(e)}, v_{j(e)})]$ to $G^\star$, wherein this weight is respect to the hyper-edge e, computed following Sect. 2. Besides, let $\mathbf{M}^\star$ denote the weighted adjacency matrix of the graph $G^\star$. Apparently, with efforts above, hyper-graph is changed into a weighted normal graph $G^\star$. With effort above, hyper-graph Laplacian can be formulated as follows: $\mathbb{L}(\mathbf{s}) = (\mathbf{I} - \mathbf{D}^{-\frac{1}{2}}\mathbf{M}^\star\mathbf{D}^{-\frac{1}{2}})\mathbf{s}$.

4.3 Hyper-Graph Convolutional Network for Short-Text Conceptualization

With efforts above, we perform GCN over this kind of simple graph G^*. Under the conventional message-passing procedure [7], when the the conventional graph convolution operation in E.q. (6), is applied to a term hyper-vertex $v \in V_T$ (representing term occurred in given short-text) in G^*, the vector representation of vertex v in the $(t + 1)$-th hidden layer, is defined as follows:

$$\mathbf{h}_v^{(t+1)} = \text{ReLU}[(\Theta^{(t)})^\top \sum_{v' \in \mathcal{N}(v)} (\overline{\mathbf{M}}^{(t)}[v][v'] \cdot \mathbf{h}_{v'}^{(t)})] \tag{7}$$

wherein, t indicates the epoch number, Θ is a matrix representing the learnable weights, $\mathcal{N}(v)$ denotes to the set of neighbours (including terms in V_T and concepts in V_C) of vertex v, and $\overline{\mathbf{M}}^{(t)}[v][v']$ is the weight on the edge (v, v') after normalisation in normalized adjacency matrix $\overline{\mathbf{M}}^{(t)}$ (respect to the adjacency matrix constructed in Sect. 2).

Because of its simplicity and its efficiency, following [32], this paper firstly only considers one representative simple edge for each hyper-edge $e \in E$ given by (v_i, v_j) where $(v_{i,e}, v_{j,e}) =: \arg\max_{v_i, v_j \in e} \|(\Theta^{(t)})^\top(\mathbf{h}_{v_i}^{(t)} - \mathbf{h}_{v_j}^{(t)})\|_l$ for epoch t. (situation of complex edge will be discussed later.) We then use Eq. (7) on each hyper-vertex $v \in V_T$, on condition of only considering such simple but representative edges incident on it, until convergence.

4.4 The Variants of the Proposed Model

There exist following three variants of the proposed hyper-graph convolutional network based model for short-text conceptualization task, with difference of prior knowledge they utilize.

hGCN-1: Leveraging Subordination Prior Knowledge. Based on the hyper-graph convolutional network (hGCN) introduced in Sect. 4.3, this paper denotes the first hGCN-based model for short-text conceptualization, as **hGCN-1**.

In **hGCN-1**, the edges in hyper-edge belong only to E_{TC}, as shown in Fig. 2 (left), because of its simplicity and its efficiency. This model mainly utilizes the subordination prior knowledge embedded in the subordination network G_{TC}, which also called "inter correlation", as discussed in Sect. 2. E.g., in Fig. 2 (left): the highlight hyper-edge consists of term hyper-vertex "apple", concept hyper-vertex COMPANY and term hyper-vertex "microsoft"; besides, only orange line, indicating whether a term belongs to a concept, exhibits. The main intuition behind **hGCN-1** is that: (i) Terms with more frequent correlations or common references to more concepts are more likely to become contextual key terms, so the concepts theys refer to together are more likely to become discriminative concepts representing this short-text; (ii) Unambiguous terminology helps to clear the surrounding ambiguous terms, which can be resolved by utilizing contextual information. In this situation, the number of the hyper-vertices embedded in the hyper-edge is no more than 3.

hGCN-2: Leveraging Concept-Correlation Prior Knowledge Based on hGCN-1. Based on the hyper-graph convolutional network (hGCN) introduced in Sect. 4.3, this paper denotes the second variant of hGCN-based model for short-text conceptualization, as **hGCN-2**.

In **hGCN-2**, the edges in hyper-edge belong to both E_{TC} and E_{CC}, as shown in Fig. 2 (middle), which additionally utilize the concept-correlation prior knowledge in the concept-correlation network G_C (described in Sect. 2), compared with **hGCN-1**. E.g., in Fig. 2 (middle): the highlight hyper-edge consists of term hyper-vertex "apple", concept hyper-vertex COMPANY, concept hyper-vertex PRODUCT and term hyper-vertex "microsoft"; besides, orange line (indicating whether a term belongs to a concept) and red line (indicating correlation relationship between concepts) exhibit. The main intuition behind **hGCN-2** is that: (i) The closer the connections between different concepts, the more likely they are to unite to represent this short-text; (ii) In the local context, influential concepts (or terms) will enhance the influence of corresponding terms (or concepts), that is, concepts and terms can be mutually strengthened through appropriate interaction. In this situation, the number of the hyper-vertices embedded in the hyper-edge is no more than 4.

hGCN-3: Leveraging Term-Correlation Prior Knowledge Based on hGCN-2. Based on the hyper-graph convolutional network (hGCN) introduced in Sect. 4.3, this paper denotes the third variant of hGCN-based model for short-text conceptualization, as **hGCN-3**.

In the most complex **hGCN-3**, the edges in hyper-edge belong to the whole $E = E_{TC} \cup E_{CC} \cup E_{TT}$, as shown in Fig. 2 (right). Compared with **hGCN-2**, it additionally leverages the term correlation prior in the term-correlation network G_T (described in Sect. 2) for modeling the interaction among terms and concepts (as well as the beneficial reactions between concepts and terms) and evaluating terms and their concepts simultaneously. E.g., in Fig. 2 (right): the highlight hyper-edge consists of term hyper-vertex "apple", concept hyper-vertex COMPANY, concept hyper-vertex PRODUCT, term hyper-vertex "ipad" and term hyper-vertex "microsoft"; besides, all kinds of line, corresponding to all kinds of correlation types, exhibit.

The main intuition behind **hGCN-3** is that: (i) There is a complementary relationship between concepts and terminology, which can be reflected in rankings; (ii) In the local context, influential concepts (or terms) will enhance the influence of corresponding terms (or concepts), that is, concepts and terms can be mutually strengthened through appropriate interaction; (iii) The fewer concepts a term involves, the stronger its discriminatory nature. In fact, terms (and concepts) interact with each other closely: Take the ambiguous term 'apple' in Fig. 1 as an example. Term 'apple' and term 'microsoft' co-occur in this context, and term 'microsoft' is unambiguous. Therefore, with the help of 'microsoft', 'apple' could be considered as concept COMPANY or BRAND, rather than concept FRUIT. In this situation, the number of the hyper-vertices embedded in the hyper-edge is no more than 5.

Variants' Comparison. As mentioned above, the significant difference among the proposed three variants of the proposed hGCN-based model for short-text conceptualization task, is the prior knowledge they utilize. The complexity of their prior knowledge increases by degrees. Apparently, as shown in Fig. 3:

(i) Only the widely-used subordination prior knowledge (also called "inter correlation") embedded in subordination network is leveraged in **hGCN-1**.
(ii) With the basic of **hGCN-1**, **hGCN-2** adds concept correlation prior knowledge embedded in the concept-correlation network G_C, into consideration.
(iii) All kinds of prior knowledge are integrated into **hGCN-3**, including subordination prior knowledge (in G_{TC}), concept correlation prior knowledge (in G_C) and term correlation prior knowledge (in G_T).

5   Experiments and Results

Due to the absence of a concept annotation corpus for short-texts, in order to validate the performance of our hyper-graph convolutional network driven model and other state-of-the-art algorithms, we manually labeled the dataset to directly

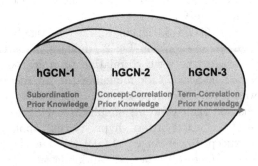

Fig. 3. Variants of the proposed hGCN-based model for short-text conceptualization with difference of prior knowledge.

evaluate concept quality, and then conducted experiments on advertising query matching tasks widely used for evaluating text conceptualization [10,20,21] to indirectly evaluate the results.

5.1 Comparative Models

The comparative models can be concluded as follows:

BOW: [17] represents short-text as bag-of-words with the TF-IDF scores.

LDA: [1] represents short-text as its inferred topic distribution, and the dimensions of the short-text vector of is number of topics as we s presuppose. In this case, topic represents "concept".

IJCAI$_{11}$: [20] proposes a probabilistic framework for short-text conceptualization task, which aims at performing a simple co-clustering of concepts and terms by identifying the disjoint cliques, and then derived the most likely concepts via Bayesian inference.

IJCAI$_{15}$: [29] conceptualizes terms using a random-walk based iterative algorithm, by taking verbs and adjectives into consideration.

Co-Rank$_{AD}$: [35] proposes a co-ranking framework by simply coupling two random walks, which separately rank different type of vertices under PageRank's paradigm [2].

Co-Rank$_{HITS}$: Hyperlink Induced Topic Search [12] is used to synergistically cope with heterogeneous information, confirming ranking could be reinforced through interactions.

HGNN: Hyper-Graph Neural Networks [6] is introduced for using the clique expansion to approximate the hyper-graph.

MLP: Multi-Layer Perceptron [13] treats each instance (i.e., hyper-vertices including the terms embedded in given short-text as well as the corresponding candidate concepts) as an independent and identically distributed (i.i.d) instance.

5.2 Experiments on Concept Quality Evaluation

In order to directly evaluate the quality of conceptualization, we construct the manually labeled dataset following [20,29]. Especially, we use Twitter's search engine to collect tweets and create a tagged tweet dataset based on keywords and topic tags, which consists of 10,500 tweets and includes commonly used ambiguous keywords. We organize 30 volunteers to label these tweets: We firstly run the comparative algorithms, and guide our volunteers to label the top-M concepts for each tweet. Wherein, notation M indicates the number of concepts of the given tweet. Given a tweet, each concept is assigned with a correlation score $rel_i = 1/0.6/0.3/0$ if it is regarded as correct/related/not-sure/unrelated to the given tweet.

Table 1. Experimental summary of concept quality evaluation.

Model	P@5	P@10
IJCAI$_{11}$ [20]	0.329	0.358
IJCAI$_{15}$ [29]	0.448	0.457
Co-Rank$_{HITS}$ [12]	0.408	0.443
Co-Rank$_{AD}$ [35]	0.439	0.456
MLP [13]	0.444	0.452
HGNN [6]	0.461^{γ}	0.460^{γ}
hGCN-1 (Ours)	0.443	0.449
hGCN-2 (Ours)	0.479	0.503
hGCN-3 (Ours)	$\mathbf{0.491}^{\gamma\psi}$	$\mathbf{0.515}^{\gamma\psi}$

Table 1 shows the experimental results of precision P@5 and precision P@10, which are evaluated by using $P@M = \sum_{i=1}^{M} rel_i/M$, respect to precision of Top-M concepts. The proposed hyper-graph convolutional network based model achieves the competitive even optimal precision on both metrics, especially with higher value of M. Specifically, our **hGCN-3** improves the P@5 and P@10 over those of **IJCAI$_{15}$** by 9.82% and 12.47%, respectively; **hGCN-3** improves the P@5 and P@10 over those of another hyper-graph neural networks based model **HGNN** by 6.72% and 11.73%, respectively.

5.3   Experiments on Ad-Query Similarity Evaluation

This section tries to introduce ads and queries matching experiment for evaluating conceptualization efficiency. As is well known, an important challenge of online advertising is to match advertising with users' information needs. As discussed in [10], the evidence of matching quality in advertising queries is whether users click on sponsored URLs in search results pages. Therefore, we look at click through rate (CTR), which is named as the number of clicks on an advertisement divided by the total number of ads placed. Therefore, if we assume that users are more likely to click on sponsored URLs that match their information needs with advertisements, then one way to improve CTR is to better match advertisements with queries. Hence, we conceptualize short-text to handle this advertising query matching task. For each sponsored URL, there exists a set of advertising keywords from the sponsor that can be kindly considered short-text. The concept vector generation method from [10], is leveraged here for releasing the sentence-level concept vector, after the proposed models and the comparative models conceptualize these short-text (i.e., ad-bid keywords). And we perform the similar operation on the user's search queries, which result in advertisements being placed in the search results. Then, we calculate the similarity between the advertising concept vector and the query concept vector. Note that, if the correlation between similarity in advertising queries and CTR is high, then we can attribute it to our successful conceptualization of short-texts. This experiment has collected Bing search logs for sponsored URLs and queries (134,261 sponsored URL query pairs from June 1, 2019 to June 30, 2019) with corresponding click through rates.

Table 2. Experimental summary of term ad-query similarity evaluation.

Model	Pearson-Correlation
BOW	0.526
LDA	0.401
IJCAI$_{11}$ [20]	0.425
IJCAI$_{15}$ [29]	0.687
Co-Rank$_{HITS}$ [12]	0.626
Co-Rank$_{AD}$ [35]	0.674
MLP [13]	0.681
HGNN [6]	0.706$^{\gamma}$
hGCN-1 (Ours)	0.698
hGCN-2 (Ours)	0.713
hGCN-3 (Ours)	**0.718**$^{\gamma\psi}$

Following the experimental settings in [10], we divide these pairs into 7 equal 0.1% boxes based on their CTR, and then aggregate pairs with CTR exceeding 0.6% into one box, and approximately 12% of the pairs fell into this aggregation box. The number of bid keywords with a CTR below 0.6% is 102,038 (76.00%). Using this dataset, we conceptualize the query and advertising keywords, and then calculate the cosine similarity between them. Form Table 2, we could conclude The proposed model (e.g., **hGCN-3**) captures the relatively high similarity for the aggregated box better than the keyword matching (such as **BOW**). The state-of-the-art probabilistic models (e.g., **IJCAI$_{11}$**) fails to show a satisfactory correlation with CTR. Take **IJCAI$_{11}$** as an example, it takes on a descending trend compared with its performance on other experimental task, even worse than basic and simple **BOW**. Moreover, it is worth noting that the performance of **hGCN-3** is better than **hGCN-1** and **hGCN-2** however not by much.

6   Conclusion

This paper presents a novel hyper-graph convolutional network driven model to solve the problem of short-text conceptualization, by leveraging hyper-graph learning strategy and prior knowledge about heterogeneous correlations

Acknowledgements. We thank anonymous reviewers for valuable comments. This work is funded by: (i) the National Natural Science Foundation of China (No. 62106243, U19B2026, U22B2601).

References

1. Blei, D.M., Ng, A.Y., Jordan, M.I.: Latent dirichlet allocation. J. Mach. Learn. Res. **3**, 993–1022 (2003)
2. Brin, S., Page, L.: The anatomy of a large-scale hypertextual web search engine. In: International Conference on World Wide Web, pp. 107–117 (1998)
3. Chan, T.H.H., Liang, Z.: Generalizing the hypergraph laplacian via a diffusion process with mediators. In: COCOON (2018)
4. Chan, T.H.H., Louis, A., Tang, Z.G., Zhang, C.: Spectral properties of hypergraph laplacian and approximation algorithms. ArXiv abs/1605.01483 (2016)
5. Chen, Z.M., Wei, X.S., Wang, P., Guo, Y.: Multi-label image recognition with graph convolutional networks. In: 2019 IEEE/CVF Conference on Computer Vision and Pattern Recognition (CVPR) (2019)
6. Feng, Y., You, H., Zhang, Z., Ji, R., Gao, Y.: Hypergraph neural networks. ArXiv abs/1809.09401 (2018)
7. Gilmer, J., Schoenholz, S.S., Riley, P.F., Vinyals, O., Dahl, G.E.: Neural message passing for quantum chemistry. ArXiv abs/1704.01212 (2017)
8. Hein, M., Setzer, S., Jost, L., Rangapuram, S.S.: The total variation on hypergraphs - learning on hypergraphs revisited. In: Advances in Neural Information Processing Systems (NIPS) (2013)
9. Huang, H., Wang, Y., Feng, C., Liu, Z., Zhou, Q.: Leveraging conceptualization for short-text embedding. IEEE Trans. Knowl. Data Eng. **30**(7), 1282–1295 (2018)
10. Kim, D., Wang, H., Oh, A.: Context-dependent conceptualization. In: International Joint Conference on Artificial Intelligence, pp. 2654–2661 (2013)
11. Kipf, T., Welling, M.: Semi-supervised classification with graph convolutional networks. ArXiv abs/1609.02907 (2016)
12. Kleinberg, J.M.: Authoritative sources in a hyperlinked environment. J. ACM **46**(5), 604–632 (1999)
13. Kruse, R., Borgelt, C., Klawonn, F., Moewes, C., Steinbrecher, M., Held, P.: Multilayer perceptrons (2013)
14. Li, P., Milenkovic, O.: Submodular hypergraphs: p-Laplacians, cheeger inequalities and spectral clustering. ArXiv abs/1803.03833 (2018)
15. Li, Z., Chen, Q., Koltun, V.: Combinatorial optimization with graph convolutional networks and guided tree search. In: Advances in Neural Information Processing Systems (NeurIPS) (2018)
16. Louis, A.: Hypergraph markov operators, eigenvalues and approximation algorithms. Proceedings of the Forty-Seventh Annual ACM Symposium on Theory of Computing (2014). https://api.semanticscholar.org/CorpusID:15089210
17. Salton, G., Mcgill, M.J.: Introduction to modern information retrieval. McGraw-Hill (1983)
18. Shi, C., Li, Y., Zhang, J., Sun, Y., Yu, P.S.: A survey of heterogeneous information network analysis. IEEE Trans. Knowl. Data Eng. **29**, 17–37 (2017)
19. Shi, J., Ji, H., Shi, C., Wang, X., Zhang, Z., Zhou, J.: Heterogeneous graph neural network for recommendation. ArXiv abs/2009.00799 (2020)
20. Song, Y., Wang, H., Wang, Z., Li, H., Chen, W.: Short text conceptualization using a probabilistic knowledgebase. In: International Joint Conference on Artificial Intelligence, pp. 2330–2336 (2011)
21. Song, Y., Wang, S., Wang, H.: Open domain short text conceptualization: a generative + descriptive modeling approach. In: International Conference on Artificial Intelligence, pp. 3820–3826 (2015)

22. Vashishth, S., Bhandari, M., Rai, P., Bhattacharyya, C., Talukdar, P.P.: Incorporating syntactic and semantic information in word embeddings using graph convolutional networks. In: ACL (2018)
23. Wang, F., Wang, Z., Li, Z., Wen, J.R.: Concept-based short text classification and ranking. In: The ACM International Conference, pp. 1069–1078 (2014)
24. Wang, Y., Huang, H., Feng, C.: Query expansion based on a feedback concept model for microblog retrieval. In: International Conference on World Wide Web, pp. 559–568 (2017)
25. Wang, Y., Huang, H., Feng, C., Zhou, Q., Gu, J., Gao, X.: CSE: conceptual sentence embeddings based on attention model. In: 54th Annual Meeting of the Association for Computational Linguistics, pp. 505–515 (2016)
26. Wang, Y., Wang, Z., Zhang, H., Liu, Z.: Microblog retrieval based on concept-enhanced pre-training model. ACM Trans. Knowl. Discov. Data **17**(3), 41:1–41:32 (2022)
27. Wang, Y., Zhang, H.: Harp: A novel hierarchical attention model for relation prediction. ACM Trans. Knowl. Discov. Data **15**, 17:1–17:22 (2021)
28. Wang, Y., Zhang, H.: Introducing graph neural networks for few-shot relation prediction in knowledge graph completion task. In: Qiu, H., Zhang, C., Fei, Z., Qiu, M., Kung, S.-Y. (eds.) KSEM 2021. LNCS (LNAI), vol. 12815, pp. 294–306. Springer, Cham (2021). https://doi.org/10.1007/978-3-030-82136-4_24
29. Wang, Z., Zhao, K., Wang, H., Meng, X., Wen, J.R.: Query understanding through knowledge-based conceptualization. In: International Conference on Artificial Intelligence, pp. 3264–3270 (2015)
30. Weston, J., Ratle, F., Mobahi, H., Collobert, R.: Deep learning via semi-supervised embedding. In: Neural Networks: Tricks of the Trade (2012)
31. Wu, W., Li, H., Wang, H., Zhu, K.Q.: Probase: a probabilistic taxonomy for text understanding. In: ACM SIGMOD International Conference on Management of Data, pp. 481–492 (2012)
32. Yadati, N., Nimishakavi, M., Yadav, P., Nitin, V., Louis, A., Talukdar, P.P.: Hyper-GCN: a new method of training graph convolutional networks on hypergraphs. In: Advances in Neural Information Processing Systems (NeurIPS 2019) (2018)
33. Zhang, C., Hu, S., Tang, Z.G., Chan, T.H.H.: Re-revisiting learning on hypergraphs: Confidence interval and subgradient method. In: International Conference on Machine Learning (ICML) (2017)
34. Zhou, D., Bousquet, O., Lal, T.N., Weston, J., Schölkopf, B.: Learning with local and global consistency. In: Advances in Neural Information Processing Systems (NIPS) (2003)
35. Zhou, D., Orshanskiy, S.A., Zha, H., Giles, C.L.: Co-ranking authors and documents in a heterogeneous network. In: 7th IEEE International Conference on Data Mining, pp. 739–744 (2007)
36. Zhu, X., Zhang, S., Zhu, Y., Zhu, P., Gao, Y.: Unsupervised spectral feature selection with dynamic hyper-graph learning. IEEE Trans. Knowl. Data Eng. **34**(6), 1–1 (2020)

What You Write Represents Your Personality: A Dual Knowledge Stream Graph Attention Network for Personality Detection

Zian Yan[1,3], Ruotong Wang[1,3], and Xiao Sun[2,3(✉)]

[1] AHU-IAI AI Joint Laboratory, Anhui University, Hefei, China
{wa21301033,wa21301031}@stu.ahu.edu.cn
[2] School of Computer Science and Information Engineering, Hefei University of Technology, Hefei, China
[3] Institute of Artificial Intelligence, Hefei Comprehensive National Science Center, Hefei, China
sunx@iai.ustc.edu.cn

Abstract. The goal of the personality detection task is to determine a person's personality traits using their social media posts. Recently, researchers have turned away from a fully data-driven approach and begun employing prior knowledge about psycholinguistic to guide their research. People typically post on social media to express their opinions or share their emotions. Therefore, it is crucial to uncover the traits and disparities in how individuals with different personalities express themselves. However, current research based on psycholinguistic principles only examines these differences superficially, failing to conduct more granular analyses, such as exploring emotions. In this paper, we propose an innovative approach that blends psycholinguistic and prior emotional knowledge to acquire features at varying levels. Our model, named Dual Knowledge Stream Graph Attention Network (DKSGAT), comprises of two streams. One stream represents posts at the psycholinguistic level, while the other encodes words at a more finely-grained emotional level based on prior emotional knowledge. Both streams' representations are then obtained to make joint inferences about personality traits. Our approach outperforms previous studies in predicting the Big Five personality and MBTI personality, as demonstrated through testing on two different public datasets.

Keywords: Personality Detection · Psycholinguistic · Emotion · Graph Attention Network · Big Five · MBTI

1 Introduction

Personality is a structure gathering behavioral, cognitive and emotional patterns [8]. Personality detection is a crucial task in numerous fields, including

F. Wu et al. (Eds.): SMP 2023, CCIS 1945, pp. 118–132, 2024.
https://doi.org/10.1007/978-981-99-7596-9_9

psychology, mental health care, and marketing [19,29], and the objective of this task is to determine a user's personality traits from online texts.

The fast evolution of social media has led to the creation of a vast number of textual posts every day, presenting the occasion for personality inference from text [20]. Earlier studies based on Linguistic Inquiry and Word Count (LIWC) [23] explore the relationship between word frequency and personality, but this way loses the context of words. With the development of the field of deep learning, deep neural networks are introduced to extract post-level features for personality detection [7,9,14]. The novel method proposed by TrigNet [30] is based on word categories in LIWC, which establish associations between user posts and acquire information from varying levels of words and posts, ultimately combining context for personality detection. Its ablation experiment reveals that the *Affect* category in LIWC has the most significant impact. Drawing inspiration from the conventional approach to identifying personality through questionnaires, PQNet [28] encodes the questions and answers from the questionnaire and employs user posts to facilitate answering the questions and therefore ascertain personality traits. A recent work [12] in Computer Vision realm used a person's long-term emotion features as input for personality detection, achieving good outcomes. While it is outside the realm of Natural Language Processing, it still serves as a personality detection task, offering the potential to identify personality through long-term emotions.

Motivated by the above observation, we develop a methodology that employs emotional value information for personality detection, proposing the **D**ual **K**nowledge **S**tream Graph **At**tention Network (DKSGAT) model for this task. Broadly, the model is composed of two streams: emotion dependency stream and psycholinguistic-emotion stream. The psycholinguistic-emotion stream is mainly an improvement from TrigNet that preserves the original network's structure and feature initialization approach. It enhances the *Affect* category by introducing more refined emotional subcategories from SenticNet7 [1] to extend it. Emotion dependency stream is a new module that we develope to capture a user's overall emotional tendency. To accomplish this, we establish dependencies between emotional words in the post and the target words of the emotional output, leveraging spaCy[1] and building a graph based on the dependencies. Then we encode the dependency graph using GAT [24] to get the final representation. Initially, BERT's [4] word embedding layer provides the word features, and all posts share the same word feature. Extensive experiments are conducted on the Essays and Kaggle datasets, and the results show that DKSGAT outperforms the baseline methods. Furthermore, we perform ablation experiments to prove the effectiveness of each module in the model.

In summary, our main contributions are as follows:

– As far as we know, this is the first time that emotions have been used for personality detection tasks, thereby providing a fresh perspective.

---

[1] https://spacy.io.

- We propose a new model called DKSGAT that is based on GAT and propose a novel method for integrating two types of external knowledge, LIWC and SenticNet7, into the model for personality detection tasks.
- Through extensive research and analysis, we demonstrate the superiority of our model over baselines as well as the effectiveness of both emotion data streams.

2   Related Work

Detecting personality from a person's online text is an emerging task in recent years and has attracted the attention of many researchers [9,14,26]. Early studies on personality detection primarily rely on manual feature engineering [16,18,31]. This involves various techniques such as psycholinguistic feature extraction via LIWC [23] or MRC [2], and statistical feature extraction through the bag-of-words model [32]. With the rapid development of deep learning, many methods using deep neural networks have achieved success in personality detection tasks. For example, [22] and [5] applied LTSM [6] on each post to predict the personality traits. [26] designed an AttRCNN module based on Inception [21] to encode each post. [9] and [7] used BERT [4] to encode each post and aggregate encoding results for classification. [14] designed a hierarchical attention network to obtain user document representations from the word granularity level to the post-granularity level in a bottom-up manner. [27] is an improvement of Transformer-XL [3], which is mainly used for multi-document classification tasks. It believed in putting together a user's posts to paint an overall personality profile without introducing post-order bias. [28] proposed a novel model to track critical information in posts with a questionnaire and provided an explicit way of identifying relevant cues in posts for personality detection, but only for MBTI personality. [30] used LIWC to establish the structure between posts at the psycholinguistic level and aggregated post information from a psychological perspective, providing a novel perspective. [33] used the psycholinguistic heterogeneous graph of TrigNet [30] and introduced self-supervised learning to extract auxiliary signals, which alleviates the over-fitting problem caused by lack of training data to a certain extent.

3   Methodology

3.1   Task Definition

Personality detection refers to a task of multi-document and multi-label classification [14,30]. Formally, given a set $P = \{p_1, p_2, \ldots, p_n\}$ of posts from a user, where $p_i = \{w_{i,1}, w_{i,2}, \ldots, w_{i,m}\}$ represents the i-th post with m words. The goal of our task is to predict T-dimensional personality traits $Y = \{y^t\}_{t=1}^T$ based on P, $T = 4$ in the MBTI personality, $T = 5$ in the Big Five personality.

3.2 Data Preprocessing

The SenticNet7 database is extensive, comprising 391,560 words and phrases. To alleviate the computational complexity, we have curated the database by eliminating the phrases, keeping only the individual words. The refined database consists of 39,510 words. The SenticNet7 assigns each word a polarity score, indicating its emotional connotation. Scores range from -1 (indicating highly negative emotion) to 1 (representing strong positive emotion). During the preprocessing phase, we compute the sentiment score of each post:

$$s_i = \frac{1}{k} \sum_{j=1}^{k} s_{i,j} \tag{1}$$

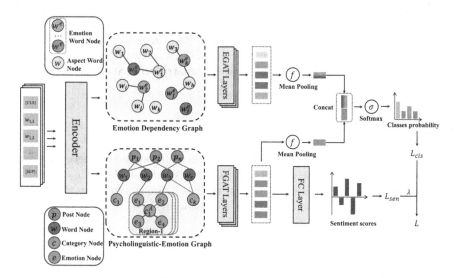

Fig. 1. An overview of our DKSGAT, which contains a emotion dependency stream(upper) to capture users global emotions and emotional bias on aspect words, and a psycholinguistic-emotion stream(lower) to enhance psycholinguistic representations with emotions.

where s_i represents the sentiment score of the entire post, and $s_{i,j}$ represents the polarity score of each word in the post that can be found in SenticNet7 database.

3.3 Emotion Dependency Stream

The upper half of the Fig. 1 illustrates the complete process of emotion dependency stream.

Graph Construction. As shown in the Fig. 1, there are two types of nodes in the emotion dependency graph $G = (V, E)$, namely *Emotion Word* and *Aspect Word*, where V denotes the set of nodes and E denotes the edges between nodes. Specifically, we define $V = V_e \cup V_a$, where $V_e = \{w_1^e, w_2^e, \ldots, w_n^e\}$ represents n emotion words that can be found in SenticNet7, and $V_a = \{w_1^a, w_2^a, \ldots, w_m^a\}$ represents m aspect words that do not appear in SenticNet but belong the relevant categories selected from LIWC. The edge $e_{i,j}$ denotes undirected connection between nodes i and j, where node i represents emotion word w_i^e and node j represents emotion word w_j^e or aspect word w_j^a. There are no edges that connect one aspect word to another.

Each user has a set of posts $P = \{p_1, p_2, \ldots, p_r\}$, and we use spaCy to parse the posts and obtain the dependency tree for each post in the Fig. 2. We identify the emotion word w_i^e in the post, and recursively parse the dependency relationships between this emotion word w_i^e and other emotion word w_j^e or aspect word w_j^a. This process results in the two types of edges("$w_i^e \rightleftharpoons w_j^e$" or "$w_i^e \rightleftharpoons w_j^a$") described above.

Our approach involves creating a graph encompassing the user's posts, whereby all posts have elements in common. Emotion word or aspect word is unique in the graph. We use a large adjacency matrix to save the edges between nodes.

Graph Initialization

Emotion Word Embedding. To capture common emotions shared between posts, we abandoned encoding the entirety of each post to obtain the node embedding of each word. Instead, we directly obtain the word embedding from the BERT[2] embedding layer. The emotion word node embeddings are represented as $X_{w^e} = \left[x_{w_1^e}, x_{w_2^e}, \ldots, x_{w_p^e} \right]^T \in \mathbb{R}^{p \times d}$.

Aspect Word Embedding. We excluded the *Function* and *Affect* categories from the LIWC that was selected by TrigNet [30]. We do not require pronouns or adverbs, etc., which are mainly found in the *Function* category. Additionally, *Affect* category belongs to LIWC's emotion category, therefore it is not needed. The aspect word node embeddings are represented as $X_{w^a} = \left[x_{w_1^a}, x_{w_2^a}, \ldots, x_{w_q^a} \right]^T \in \mathbb{R}^{q \times d}$. All nodes of emotion dependency graph are represented as $X_w = X_{w^e} \cup X_{w^a}$, where d is the dimension of each representation.

---

[2] We use the "bert-base-uncased" version.

Fig. 2. An example of using spaCy to analyze a post to obtain the dependency tree between emotion words and aspect words.

Graph Learning. Inspired by HSG [25] and TrigNet, we use GAT to compute weighted attention scores and update node representations based on the edges between nodes, and we add two flows "$w^e \leftrightarrow w^a \leftrightarrow w^e$" and "$w^e \leftrightarrow w^e$" on the original GAT model for emothion dependency graph. The flow means information interaction between left and right nodes. Specifically, the $(l+1)$-th layer emotion flow GAT updates the word node embeddings as follows:

$$\mathbf{H_w}^{(l+1)} = \text{EGAT}\left(\mathbf{H_w}^{(l)}\right) \tag{2}$$

where $\mathbf{H_w}^{(1)} = X_w$. The EGAT $(\cdot)$ contains two flows:

$$\begin{aligned}
\mathbf{H}^{(l)}_{\mathbf{w^a} \leftarrow \mathbf{w^e}} &= \text{EMP}\left(\mathbf{H}^{(l)}_{\mathbf{w^a}}, \mathbf{H}^{(l)}_{\mathbf{w^e}}\right) \\
\mathbf{H}^{(l)}_{\mathbf{w^e} \leftarrow \mathbf{w^a}, \mathbf{w^e}} &= \text{EMP}\left(\mathbf{H}^{(l)}_{\mathbf{w^e}}, \mathbf{H}^{(l)}_{\mathbf{w^a} \leftarrow \mathbf{w^e}}\right)
\end{aligned} \tag{3}$$

$$\mathbf{H}^{(l)}_{\mathbf{w^e} \leftarrow \mathbf{w^e}, \mathbf{w^a}, \mathbf{w^e}} = \text{EMP}\left(\mathbf{H}^{(l)}_{\mathbf{w^e}}, \mathbf{H}^{(l)}_{\mathbf{w^e} \leftarrow \mathbf{w^a}, \mathbf{w^e}}\right) \tag{4}$$

$$\mathbf{H}^{(l+1)}_{\mathbf{w^e}} = \text{mean}\left(\mathbf{H}^{(l)}_{\mathbf{w^e} \leftarrow \mathbf{w^a}, \mathbf{w^e}}, \mathbf{H}^{(l)}_{\mathbf{w^e} \leftarrow \mathbf{w^e}, \mathbf{w^a}, \mathbf{w^e}}\right) \tag{5}$$

where $\leftarrow$ means the direction of information movement is from right nodes to left nodes, mean$(\cdot)$ denotes the mean pooling function, and EMP$(\cdot)$ denotes the emotion information movement funciton. Equation(3) and Eq.(4) are concrete implementations of two flows "$w^e \leftrightarrow w^a \leftrightarrow w^e$" and "$w^e \leftrightarrow w^e$".

Actually, the EMP$(\cdot)$ is implemented by multi-head attention. The EMP$(\cdot)$ function's first parameter is used as query, and the second parameter is used as key and value. Next, we take EMP$\left(\mathbf{H}^{(l)}_{\mathbf{w^a}}, \mathbf{H}^{(l)}_{\mathbf{w^e}}\right)$ (the first line of Eq.(3)) as an example to illustrate the calculation process of emotion flow GAT, where $\mathbf{H}^{(l)}_{\mathbf{w^a}} = \left[h^{(l)}_{w^a_1}, h^{(l)}_{w^a_2}, \ldots, h^{(l)}_{w^a_q}\right]$, and $\mathbf{H}^{(l)}_{\mathbf{w^e}} = \left[h^{(l)}_{w^e_1}, h^{(l)}_{w^e_2}, \ldots, h^{(l)}_{w^e_p}\right]$. First, we calculate the attention weight score α^u_{ij} between node i in V_a and its neightbor node j in V_e:

$$z^u_{ij} = \text{LeakyReLu}\left(\mathbf{W}^u_{\mathbf{z}}\left[\mathbf{W}^u_{\mathbf{q}}h^{(l)}_{w^a_i} \| \mathbf{W}^u_{\mathbf{k}}h^{(l)}_{w^e_j}\right]\right) \tag{6}$$

$$\alpha^u_{ij} = \frac{\exp\left(z^u_{ij}\right)}{\sum_{r \in \mathcal{N}_i} \exp\left(z^u_{ir}\right)} \tag{7}$$

where u denotes the u-th head, $\mathbf{W}_{\mathbf{q}}^{u}$, $\mathbf{W}_{\mathbf{k}}^{u}$ and $\mathbf{W}_{\mathbf{z}}^{u}$ are learnable weight matrixes, $\mathcal{N}_i$ denotes the neighbor nodes of node i, and $\|$ is the concatenation operation. Then, we update the hidden representation:

$$\hat{h}_{w_i^a}^{(l)} = \|_{u=1}^{U} \sigma \left(\sum_{j \in \mathcal{N}_i} \alpha_{ij}^u \mathbf{W}_{\mathbf{v}}^u h_{w_j^e}^{(l)} \right) \tag{8}$$

where U is the number of heads, σ is the tanh activation function, and $\mathbf{W}_{\mathbf{v}}$ is a learnable weight martrix.

3.4 Psycholinguistic-Emotion Stream

The lower half of the Fig. 1 illustrates the complete process of psycholinguistic-emotion stream. The psycholinguistic-emotion graph is improvement to TrigNet's psycholinguistic graph, and the FGAT is improvement to the flow GAT. The succeeding subsections provide a detailed description of the specific methods for improvement.

Graph Augmentation. We use the psycholinguistic graph from the TrigNet and augment it with SenticNet7, called psycholinguistic-emotion graph. Our observations indicate that the lack of *Affect* category has the most substantial impact on TrigNet's model prediction accuracy, surpassing other categories in its ablation study. The *Affect* category of LIWC includes emotion subcategories; however, it only features basic subcategories that differentiate between positive and negative emotions, displaying a low resolution. Consequently, the SeticNet7 database is used to enhance it with additional emotion subcategories. Notably, all of the newly appended subcategories belong to *Affect* and are retrievable in LIWC. Specifically, we introduce a set of emotion categories $E = \{e_1, e_2, \ldots, e_{24}\}$, which includes 24 categories of emotions, encompassing all emotion categories in SeticNet7. Specifically, We calculate the polarity scores of all words within each category, and then calculate their average to determine the final polarity score for that particular category. According to the polarity score of each subcategory, the similar subcategories are divided into one region, and the specific division results are as follows:

- Region-1: Delight, Enthusiasm, Bliss, Ecstasy.
- Region-2: Pleasantness, Joy, Eagerness, Calmness.
- Region-3: Acceptance, Responsiveness, Contentment, Serenity.
- Region-4: Melancholy, Anxiety, Annoyance, Dislike.
- Region-5: Fear, Anger, Sadness, Disgust.
- Region-6: Terror, Grief, Loathing, Rage.

We replace the "UNUSED" tokens in BERT's vocabulary with the 24 category names, using a lookup in the BERT embedding layer to generate their embeddings $X_e = \left[x_{e_1}, x_{e_2}, \ldots, x_{e_y} \right]^T \in \mathbb{R}^{y \times d}$. Additionally, we allocate a center node, denoted as c^A, to each region. These center nodes possess identical initialization weights but are not weight-shared. For example, for each emotion subcategory node e_i from Region-1, there is an undirected edge connecting it to the c_1^A node.

Emotion Flow. Following the acquisition of the psycholinguistic-emotion graph, we add a novel data flow method, known as the emotion flow, based on TrgiNet's flow GAT to more effectively capture emotional correlations amongst posts. The emotion flow between posts in the graph is implemented by "$p \leftrightarrow w \leftrightarrow e \leftrightarrow c^A \leftrightarrow e \leftrightarrow w \leftrightarrow p$", which means that posts interact by emotive words that have the similar emotion categories (e.g., "$p_1 \leftrightarrow w_1 \leftrightarrow e_1 \leftrightarrow c^A \leftrightarrow e_2 \leftrightarrow w_2 \leftrightarrow p_2$"):

$$\mathbf{H}_{e \leftarrow w,p}^{(l)} = \mathrm{MP}\left(\mathbf{H}_e^{(l)}, \mathbf{H}_{w \leftarrow p}^{(l)}\right)$$

$$\mathbf{H}_{c^A \leftarrow e,w,p}^{(l)} = \mathrm{MP}\left(\mathbf{H}_{c^A}^{(l)}, \mathbf{H}_{e \leftarrow w,p}^{(l)}\right)$$

$$\mathbf{H}_{e \leftarrow c^A,e,w,p}^{(l)} = \mathrm{MP}\left(\mathbf{H}_{e \leftarrow w,p}^{(l)}, \mathbf{H}_{c^A \leftarrow e,w,p}^{(l)}\right) \tag{9}$$

$$\mathbf{H}_{w \leftarrow e,c^A,e,w,p}^{(l)} = \mathrm{MP}\left(\mathbf{H}_{w \leftarrow p}^{(l)}, \mathbf{H}_{e \leftarrow c^A,e,w,p}^{(l)}\right)$$

$$\mathbf{H}_{p \leftarrow w,e,c^A,e,w,p}^{(l)} = \mathrm{MP}\left(\mathbf{H}_p^{(l)}, \mathbf{H}_{w \leftarrow e,c^A,e,w,p}^{(l)}\right)$$

where $\mathbf{H}_p^{(l)}$ and $\mathbf{H}_{w \leftarrow p}^{(l)}$ are the results of original flow, $\mathbf{H}_e^{(l)}$ is the hidden state of E at the l-th layer, $\mathbf{H}_e^{(1)} = X_e$, $\mathrm{MP}(\cdot)$ denotes the message passing function in TrigNet's flow GAT, and $\mathbf{H}_{p \leftarrow w,e,c^A,e,w,p}^{(l)}$ is also used as the third part of calculating $\mathbf{H}_p^{(l+1)}$.

Auxiliary Task. We augment the psycholinguistic-emotion stream with an auxiliary task aimed at predicting the sentiment score for each post, which falls under the regression task. Formally, after encoding a set of posts $P = \{p_1, p_2, \ldots, p_n\}$ using the augmented flow GAT, the resultant representation $H_p = [h_{p_1}, h_{p_1}, \ldots, h_{p_n}]^T \in \mathbb{R}^{n \times d}$ is utilized to compute the sentiment scores for each post:

$$\hat{s}_{p_i} = \sigma\left(\mathrm{FC}(h_{p_i})\right) \tag{10}$$

where $\mathrm{FC}(\cdot)$ is fully connected layer, and σ is the tanh activation function. The real label value is obtained according to the calculation of the data preprocessing section(3.2). Ultimately, the objective function for the auxiliary task of predicting sentiment scores is expressed as follow:

$$L_{sen} = \frac{1}{T} \sum_{t=1}^{T} \sum_{i=1}^{N} (\hat{s}_{p_i} - s_{p_i})^2 \tag{11}$$

where T is the number of training samples, N is the number of posts from a sample, $\hat{s}_{p_i}$ is the predicted score for the i-th post, and s_{p_i} is the true label value.

3.5   Classification and Objective

Through emotion dependency stream and psycholinguistic-emotion stream, we obtain emotion words in all posts of a user representation $H_{w^e} = \left[h_{w_1^e}, h_{w_2^e}, \ldots, h_{w_m^e}\right]$ and all posts representation $H_p = [h_{p_1}, h_{p_2}, \ldots, h_{p_n}]$. By using the mean pooling function, we obtain the final representations of the two streams $\mathbf{h}_{w^e}$ and $\mathbf{h}_p$:

$$
\begin{aligned}
\mathbf{h}_{w^e} &= \text{mean}\left(\left[h_{w_1^e}, h_{w_2^e}, \ldots h_{w_m^e}\right]\right) \\
\mathbf{h}_p &= \text{mean}\left[\left(h_{p_1}, h_{p_2}, \ldots h_{p_n}\right)\right]
\end{aligned} \tag{12}
$$

We concatenate $\mathbf{h}_{w^e}$ and $\mathbf{h}_p$ as the final representation for classification task:

$$
u = \text{concat}\left(\mathbf{h}_{w^e}, \mathbf{h}_p\right) \tag{13}
$$

Then, we predict T personality traits by using T softmax-normalized linear transformations. For the t-th personality trait, we calculate:

$$
\hat{y}_t = \text{softmax}\left(u\mathbf{W}_u^t + \mathbf{b}_u^t\right) \tag{14}
$$

where $\mathbf{W}_u^t$ is a trainable weight matrix and $\mathbf{b}_u^t$ is a bias part. The objective function of classification task defined as:

$$
L_{cls} = -\frac{1}{U}\sum_{u}^{U}\sum_{t=1}^{T}\left(y_u^t \log\left(\hat{y}_u^t\right) + \left(1 - y_u^t\right)\log\left(1 - \hat{y}_u^t\right)\right) \tag{15}
$$

where U is the number of training samples, T is the number of personality traits, y_u^t is the true label of the t-th trait, and $\hat{y}_u^t$ is the predicted value. Finally, we concurrently train the personality detection classification task and the sentiment score prediction task using a joint objective function:

$$
L = L_{cls} + \lambda \cdot L_{sen} \tag{16}
$$

where λ is a trade-off parameter.

4   Experiments

4.1   Datasets

We use two datasets Essays [17] and Kaggle[3] to verify the effectiveness of our model, and divide the datasets into training set, validation set and test set in a ratio of 6:2:2.

4.2   Baselines

We employ the following models as baselines to compare with our model: BiL-STM [22], BERT [9], RoBERTa [13], AttRCNN [26], SN+Attn [14], Transformer-MD [27], TrigNet [30] and PQNet [28].

---

[3] https://www.kaggle.com/datasnaek/mbti-type.

4.3   Implementation Details

We use Pytorch[4] to implement our model and train it on four NVIDIA GeForce RTX 3090 GPUs. Following previous work TrigNet [30], we use Adam [10] as optimizer, and set the learning rate of BERT to 2e-5, and 1e-3 to other parts. For the Essays dataset, the maximum number of nodes in the graph is limited to 500, while for the Kaggle dataset, it is increased to 530 owing to the addition of 24 new subcategories and 6 center nodes. The statistics reveal that, in total, less than 2% of the samples in the two datasets exceed 500 nodes. After tuning on the validation dataset, we detect an overfitting issue with the model and resolve it by augmenting the dropout probability from 0.2 (used in the TrigNet model) to 0.5. The training mini-batch size is set to 16. For the auxiliary task, the value of λ is searched in $\{0.01, 0.05, 0.1, 0.2, 0.3, 0.4, 0.5, 1.0\}$, and the specific result is shown in the Fig. 4.

5   Analysis

Table 1. Overall results of different models in Macro-F1(%), where the best results are shown in bold. The PQNet model can only be used for the detection of MBTI personality. The DKSGAT/GCN refers to the substitution of GCN for GAT in emotion dependency stream.

Methods	Essays						Kaggle				
	OPN	CON	EXT	AGR	NEU	Average	I/E	S/N	T/F	P/J	Average
BiLSTM	63.32	62.47	63.54	65.97	56.30	62.32	57.82	57.87	69.97	57.01	60.67
BERT_finetune	65.13	64.55	67.12	68.14	60.51	65.09	64.65	57.12	77.95	65.25	66.24
RoBERTa_finetune	65.76	66.03	66.91	68.38	60.45	65.51	64.47	59.23	78.11	65.82	66.91
AttRCNN	67.84	63.46	71.50	71.92	62.36	67.42	59.74	64.08	78.77	66.44	67.25
SN+Attn	68.50	64.19	**72.25**	70.82	68.10	68.77	65.43	62.15	78.05	63.92	67.39
Transformer-MD	69.63	68.14	71.89	70.47	69.45	69.92	66.08	**69.10**	79.19	67.50	70.47
TrigNet	69.52	68.27	70.01	73.12	69.34	70.05	69.54	67.17	79.06	67.69	70.86
PQNet	–	–	–	–	–	–	68.94	67.65	79.12	**69.57**	71.32
DKSGAT/GCN	68.46	67.61	70.24	72.93	**70.15**	69.88	68.48	68.65	78.13	68.34	70.90
DKSGAT	**70.87**	**68.82**	71.51	**74.28**	69.91	**71.08**	**70.13**	66.87	**80.41**	69.50	**71.73**

5.1   Overall Results

The overall results are showed in Table 1. The proposed model DKSGAT outperforms the state-of-the-art model in terms of average F1 score on both Essays and Kaggle datasets. We attempt to substitute GCN [11] for GAT in emotion dependency stream. However, the outcome is suboptimal.

---

[4] https://pytorch.org/.

5.2   Ablation Study

To further investigate the impact of each key module, we perform ablation studies by sequentially deleting them from DKSGAT. The results are divided into three groups, as shown in the Table 2.

In the first group of ablation experiments, we investigate the contribution of the main modules in the model. The removal of emotion dependency stream result in a decrease of 0.96 and 1.04 in the Marcro-F1 scores of the model in Essays and Kaggle, respectively.

Table 2. Results of ablation study in average Macro-F1 (%) on the Essays and Kaggle datasets, where "w/o" means removal of a component from the original DKSGAT, and "Δ" represents the performance change.

Methods	Essays		Kaggle	
	Macro-F1	(Δ %)	Macro-F1	(Δ %)
DKSGAT	**71.08**	–	**71.73**	–
w/o emotion dependency stream	70.12	0.96↓	70.69	1.04↓
w/o psycholinguistic-emotion stream	69.34	1.74↓	69.52	2.21↓
w/o auxiliary task	70.43	0.65↓	70.96	0.77↓
w/o "$w^e \leftrightarrow w^a \leftrightarrow w^e$"	70.31	0.77↓	70.89	0.84↓
w/o "$w^e \leftrightarrow w^e$"	70.60	0.48↓	71.22	0.51↓
w/o "$p \leftrightarrow w \leftrightarrow e \leftrightarrow c^A \leftrightarrow e \leftrightarrow w \leftrightarrow p$"	69.89	1.19↓	70.90	0.83↓
w/o Region-1	70.07	1.01↓	70.79	0.94↓
w/o Region-2	70.69	0.39↓	71.25	0.48↓
w/o Region-3	70.63	0.45↓	71.19	0.54↓
w/o Region-4	70.81	0.27↓	71.52	0.21↓
w/o Region-5	70.89	0.19↓	71.48	0.25↓
w/o Region-6	70.46	0.62↓	71.00	0.73↓

In the second group, we can observe that after removing "$w^e \leftrightarrow w^a \leftrightarrow w^e$" and "$w^e \leftrightarrow w^e$", both have different degrees of impact on the model performance. The lack of "$p \leftrightarrow w \leftrightarrow e \leftrightarrow c^A \leftrightarrow e \leftrightarrow w \leftrightarrow p$" means that we have not augmented the flow GAT, which results in unequal emotional information in the two streams.

In the third group, the model performance is most influenced by Region-1 and Region-6, followed by Region-2 and Region-3, and the impact of Region-4 and Region-5 is relatively small. Our statistics show that the impact is directly proportional to the word count in Region, i.e., greater number of words lead to greater impact.

5.3   Visualization Analysis

To further illustrate that the representations learned by emotion dependency stream are inductive and distinguishable, we use t-SNE [15] to reduce the dimension of the learned representations in the T/F trait of Kaggle test set to 2 and visualize the effect. As the results in Fig. 3 show, the representation learned by emotion dependency stream exhibits clear discrimination such that each class features distinct cluster centers and a relatively compact distribution of features.

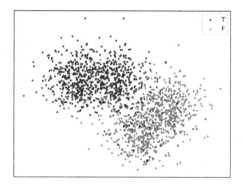

Fig. 3. Visualization result of the produced user representations by emotion dependency stream.

5.4   Effect of Trade-Off Parameter

Figure 4 shows the change trend of average Macro-F1 score of DKSGAT model when the parameter λ increases. We can observe that the score initially increases as the trade-off parameter λ increases, however, it starts to decline once λ exceeds 0.1. The model's performance markedly declines when lambda exceeds 0.3.

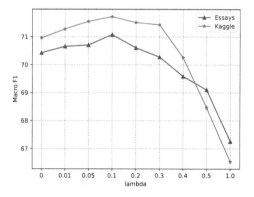

Fig. 4. Performance curves for different value of lambda.

6 Conclusion

In this paper, we introduce a novel approach to enhancing the association between posts, namely, incorporating emotional information. We propose a model called DKSGAT to achieve this objective. We attempt to introduce finer-grained prior knowledge in the context of psycholinguistic to select posts strongly associated with personality. To sum up, on the data-scarce task of personality detection, only relying on data-driven methods has a lower upper limit, and future studies should consider other personality-related factors to maximize the potential of the detection process.

Acknowledgment. This work was supported by the National Key R&D Programme of China (2022YFC3803202), Major Project of Anhui Province under Grant 202203a05020011.

References

1. Cambria, E., Liu, Q., Decherchi, S., Xing, F., Kwok, K.: SenticNet 7: a commonsense-based Neurosymbolic AI framework for explainable sentiment analysis. In: Proceedings of the Thirteenth Language Resources and Evaluation Conference, pp. 3829–3839 (2022)
2. Coltheart, M.: The MRC psycholinguistic database. Q. J. Exp. Psychol. A **33**(4), 497–505 (1981)
3. Dai, Z., Yang, Z., Yang, Y., Carbonell, J., Le, Q.V., Salakhutdinov, R.: Transformer-XL: Attentive language models beyond a fixed-length context. arXiv preprint arXiv:1901.02860 (2019)
4. Devlin, J., Chang, M.W., Lee, K., Toutanova, K.: BERT: Pre-training of deep bidirectional transformers for language understanding. arXiv preprint arXiv:1810.04805 (2018)
5. Hernandez, R., Scott, I.: Predicting Myers-Briggs type indicator with text. In: 31st Conference on Neural Information Processing Systems (NIPS 2017) (2017)
6. Hochreiter, S., Schmidhuber, J.: Long short-term memory. Neural Comput. **9**(8), 1735–1780 (1997)
7. Jiang, H., Zhang, X., Choi, J.D.: Automatic text-based personality recognition on monologues and multiparty dialogues using attentive networks and contextual embeddings (student abstract). In: Proceedings of the AAAI Conference on Artificial Intelligence. vol. 34, pp. 13821–13822 (2020)
8. Kaushal, V., Patwardhan, M.: Emerging trends in personality identification using online social networks–a literature survey. ACM Trans. Knowl. Disc. Data (TKDD) **12**(2), 1–30 (2018)
9. Keh, S.S., Cheng, I., et al.: Myers-Briggs personality classification and personality-specific language generation using pre-trained language models. arXiv preprint arXiv:1907.06333 (2019)
10. Kingma, D.P., Ba, J.: Adam: A method for stochastic optimization. arXiv preprint arXiv:1412.6980 (2014)
11. Kipf, T.N., Welling, M.: Semi-supervised classification with graph convolutional networks. arXiv preprint arXiv:1609.02907 (2016)

12. Liu, F., et al.: OPO-FCM: a computational affection based OCC-PAD-OCEAN federation cognitive modeling approach. IEEE Trans. Comput. Soc. Syst. **10**(4), 1813–1825 (2022)
13. Liu, Y., et al.: RoBERTa: A robustly optimized BERT pretraining approach. arXiv preprint arXiv:1907.11692 (2019)
14. Lynn, V., Balasubramanian, N., Schwartz, H.A.: Hierarchical modeling for user personality prediction: the role of message-level attention. In: Proceedings of the 58th Annual Meeting of the Association for Computational Linguistics, pp. 5306–5316 (2020)
15. Van der Maaten, L., Hinton, G.: Visualizing data using t-SNE. J. Mach. Learn. Res. **9**(11), 2579–2605 (2008)
16. Mairesse, F., Walker, M.A., Mehl, M.R., Moore, R.K.: Using linguistic cues for the automatic recognition of personality in conversation and text. J. Artif. Intell. Res. **30**, 457–500 (2007)
17. Pennebaker, J.W., King, L.A.: Linguistic styles: language use as an individual difference. J. Pers. Soc. Psychol. **77**(6), 1296 (1999)
18. Schwartz, H.A., et al.: Personality, gender, and age in the language of social media: the open-vocabulary approach. PLoS ONE **8**(9), e73791 (2013)
19. Shen, T., et al.: PEIA: personality and emotion integrated attentive model for music recommendation on social media platforms. In: Proceedings of the AAAI Conference on Artificial Intelligence. vol. 34, pp. 206–213 (2020)
20. Štajner, S., Yenikent, S.: A survey of automatic personality detection from texts. In: Proceedings of the 28th International Conference on Computational Linguistics, pp. 6284–6295 (2020)
21. Szegedy, C., Ioffe, S., Vanhoucke, V., Alemi, A.: Inception-v4, inception-resNet and the impact of residual connections on learning. In: Proceedings of the AAAI Conference on Artificial Intelligence. vol. 31 (2017)
22. Tandera, T., Suhartono, D., Wongso, R., Prasetio, Y.L., et al.: Personality prediction system from Facebook users. Procedia Comput. Sci. **116**, 604–611 (2017)
23. Tausczik, Y.R., Pennebaker, J.W.: The psychological meaning of words: LIWC and computerized text analysis methods. J. Lang. Soc. Psychol. **29**(1), 24–54 (2010)
24. Veličković, P., Cucurull, G., Casanova, A., Romero, A., Lió, P., Bengio, Y.: Graph attention networks (2018)
25. Wang, D., Liu, P., Zheng, Y., Qiu, X., Huang, X.: Heterogeneous graph neural networks for extractive document summarization. arXiv preprint arXiv:2004.12393 (2020)
26. Xue, D., et al.: Deep learning-based personality recognition from text posts of online social networks. Appl. Intell. **48**, 4232–4246 (2018)
27. Yang, F., Quan, X., Yang, Y., Yu, J.: Multi-document transformer for personality detection. In: Proceedings of the AAAI Conference on Artificial Intelligence, vol. 35, pp. 14221–14229 (2021)
28. Yang, F., Yang, T., Quan, X., Su, Q.: Learning to answer psychological questionnaire for personality detection. In: Findings of the Association for Computational Linguistics: EMNLP 2021, pp. 1131–1142 (2021)
29. Yang, H.C., Huang, Z.R.: Mining personality traits from social messages for game recommender systems. Knowl.-Based Syst. **165**, 157–168 (2019)
30. Yang, T., Yang, F., Ouyang, H., Quan, X.: Psycholinguistic tripartite graph network for personality detection. arXiv preprint arXiv:2106.04963 (2021)
31. Yarkoni, T.: Personality in 100,000 words: a large-scale analysis of personality and word use among bloggers. J. Res. Pers. **44**(3), 363–373 (2010)

32. Zhang, Y., Jin, R., Zhou, Z.H.: Understanding bag-of-words model: a statistical framework. Int. J. Mach. Learn. Cybern. **1**, 43–52 (2010)
33. Zhu, Y., Hu, L., Ge, X., Peng, W., Wu, B.: Contrastive graph transformer network for personality detection. In: Proceedings of the Thirty-First International Joint Conference on Artificial Intelligence (2022)

Detect Depression from Social Networks with Sentiment Knowledge Sharing

Yan Shi[1], Yao Tian[1], Chengwei Tong[1], Chunyan Zhu[2,3], Qianqian Li[3], Mengzhu Zhang[2], Wei Zhao[4], Yong Liao[1], and Pengyuan Zhou[1(✉)]

[1] University of Science and Technology of China, Hefei, China
{syan0724,tyzkd,cwtong}@mail.ustc.edu.cn, {yliao,pyzhou}@ustc.edu.cn
[2] Anhui Medical University, Hefei, China
[3] The Second Affiliated Hospital of Anhui Medical University, Hefei, China
[4] Anhui ERCIASII, Anhui University of Technology, Ma'anshan, China
zhaowei@ahut.edu.cn

Abstract. Social network plays an important role in propagating people's viewpoints, emotions, thoughts, and fears. Notably, following lockdown periods during the COVID-19 pandemic, the issue of depression has garnered increasing attention, with a significant portion of individuals resorting to social networks as an outlet for expressing emotions. Using deep learning techniques to discern potential signs of depression from social network messages facilitates the early identification of mental health conditions. Current efforts in detecting depression through social networks typically rely solely on analyzing the textual content, overlooking other potential information. In this work, we conduct a thorough investigation that unveils a strong correlation between depression and negative emotional states. The integration of such associations as external knowledge can provide valuable insights for detecting depression. Accordingly, we propose a multi-task training framework, DeSK, which utilizes shared sentiment knowledge to enhance the efficacy of depression detection. Experiments conducted on both Chinese and English datasets demonstrate the cross-lingual effectiveness of DeSK.

Keywords: Social networks · Depression detection · Sentiment analysis · Multi-task learning

1 Introduction

In recent years, there has been a growing focus on the subject of mental health, drawing significant attention from the general public. Particularly following the outbreak of the COVID-19 pandemic, a surge in the prevalence of common mental health disorders has been observed [2]. And among these disorders, depression stands out as the most prevalent form, exhibiting a strong correlation with substantial morbidity and mortality rates [4]. Traditional methods for depression diagnosis usually use interviews with patients or self-report questionnaires, which are time-consuming and error-prone.

F. Wu et al. (Eds.): SMP 2023, CCIS 1945, pp. 133–146, 2024.
https://doi.org/10.1007/978-981-99-7596-9_10

The social network offers a pathway for capturing pertinent behavioral attributes pertaining to an individual's cognition, emotional state, communication patterns, daily activities, and social interactions. The emotions conveyed and the linguistic patterns employed in posts on social networks can potentially serve as indicators of prevailing sentiments such as feelings of insignificance, culpability, powerlessness, and intense self-disdain, which are characteristic of major depressive disorder [5]. Hence, it becomes paramount to comprehend and analyze the emotions that individuals convey through social networks, especially during difficult times like the pandemic. Furthermore, the timely identification of initial indicators of depression is of great importance, as it enables prompt intervention and assistance for those in need.

There is a large amount of existing work analyzing depression using social network data. Early works are usually based on statistical or traditional machine-learning approaches. [20] proposed a multi-modal depressive dictionary learning model specifically for detecting users with depressive tendencies on Twitter. Recently, deep learning methods have exhibited remarkable advancements, attaining notable levels of performance in depression detection. [3] proposed a deep learning model named X-A-BiLSTM for depression detection in imbalanced social network data. Leveraging the capabilities of the Transformer model, [12] demonstrated significantly improved accuracy in detecting depression among social network users.

Nevertheless, the existing methods predominantly focus on using pre-trained models or deeper networks to obtain the semantic aspects of sentences. The neglect of sentiment features of the target sentences and external sentiment knowledge leads to unsatisfactory performance [1,24] of the neural network in depression detection. Because depression is usually highly correlated with negative emotions, these sentiment features could contribute to more comprehensive and accurate depression detection. With the intuition that sentiment serves as a direct clue to depression [17], we introduce external sentiment knowledge into depression detection to enhance performance. Specifically, our approach incorporates external sentiment knowledge into the depression detection model by leveraging a multi-task learning framework. This framework facilitates the simultaneous learning of sentiment analysis and depression detection, enabling the model to benefit from the additional information provided by external sentiment knowledge. The main contributions of this work are summarised as follows:

1. We propose a Depression detection based on Sentiment Knowledge model called **DeSK**, which employs multi-task learning to acquire and leverage external sentiment knowledge which is overlooked by previous works.
2. Considering the scarcity of publicly available datasets pertaining to depression in Chinese social networks, we collect and construct a dataset focused on depression from the Weibo platform. This dataset was created through self-diagnosis methods for further research and analysis[1].

---

[1] Contact the second author to inquire about the dataset.

3. Experimental results on the Reddit dataset show that DeSK outperforms state-of-the-art performance. The ablation tests validate the model components and demonstrate their efficacy in detecting depression.

2   Related Work

Depression detection and sentiment analysis have been extensively studied in the field of mental health analysis using social network data. Early approaches focused on utilizing statistical or traditional machine learning methods to detect depression based on textual content. These methods mainly relied on linguistic patterns and semantic information to identify signs of depression. However, they often overlooked the crucial role of sentiment features in depression detection. However, multi-task learning offers a promising solution to address this disadvantage. By jointly learning these tasks, the models can benefit from shared knowledge and improve performance in both domains.

2.1   Depression Detection

Trained professional psychologists rely on various methods, including written descriptions provided by individuals and psychometric assessments, to assess and diagnose depression accurately [16]. Social network-based sentiment analysis is an alternative depression detection approach rising in recent years. Researchers can extract valuable insights from the vast amount of data available on social networks, such as patterns, trends, and user-generated content related to depression.

By extracting various behavioral attributes from social network platforms, such as social engagement, mood, speech and language style, self-networking, and mentions of antidepressants, [5] aimed to provide estimates of depression risk. [3] proposed a deep learning model (X-A-BiLSTM) to handle the real-world imbalanced data distributions in social networks for depression detection. [9] proposed a deep visual-textual multi-modal learning approach aimed at acquiring robust features from both normal users and users diagnosed with depression. And [8] developed a depression lexicon based on domain knowledge of depression to facilitate better extraction of lexical features related to depression.

2.2   Sentiment Analysis

Sentiment analysis seeks to examine individuals' sentiments or opinions concerning various entities, including but not limited to topics, events, individuals, issues, services, products, organizations, and their associated attributes [29]. For the past few years, the growth of social networks has significantly propelled the advancement of sentiment analysis. To date, the majority of sentiment analysis research is based on natural language processing techniques. [7] combined text features and machine learning methods to classify social network texts into six types of emotions. [21] employed machine learning algorithms, specifically Naive Bayes (NB) and the k-nearest neighbor algorithm (KNN), to discern the emotional content of Twitter messages and classified the Twitter messages into four distinct emotional categories.

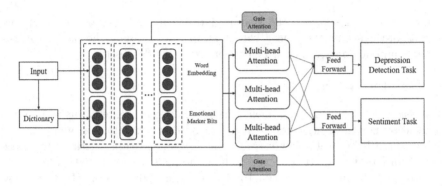

Fig. 1. The overall framework of DeSK. (Color figure online)

2.3   Multi-Task Learning

Multi-Task Learning (MTL) is a machine learning paradigm that aims to enhance the generalization performance of multiple related tasks by leveraging the valuable information inherent in these tasks [30]. [31] proposed a facial landmark detection by combining head pose estimation and facial attribute inference. [32] incorporated sentiment knowledge into a hate speech detection task by employing a multi-task learning framework.

3   Methodology

In this section, we introduce DeSK, which exhibits an enhanced capability to detect depression through the integration of target sentence sentiment and external sentiment knowledge. The overall architecture is shown in Fig. 1. The framework primarily comprises three components: 1) Input layer, which captures the sentiment features of the sentence. A depressed words dictionary is employed to determine if each word exhibits depressed speech characteristics and appended to the word embedding as emotional marker bits. 2) Multi-task learning framework, which leverages the strong correlation between sentiment analysis and depression detection, to model task relationships and acquire task-specific features by leveraging shared sentiment knowledge. Multiple feature extraction units, consisting of a multi-head attention layer and a feed-forward neural network, are utilized for this purpose. 3) Gated attention layer, which is a gated attention mechanism that calculates the probability of selecting each feature extraction unit.

3.1   Input Layer

The central idea of DeSK revolves around the notable association between depression and negative emotions. We hypothesize that texts expressing depressive sentiments frequently contain explicit usage of negative emotion words [23].

Hence, directing attention towards capturing derogatory words within a sentence can assist in enhancing depression detection capabilities. More specifically, we capture this sentiment information by utilizing sentiment marker bits.

Sentiment Marker Bits. Our work is based on the intuition that certain specific words that possess an exceedingly negative nature, such as sadness [15], disgust [28], etc., have a more substantial impact on the assessment of depression. To address this, we have constructed a depressed word dictionary, the vocabulary of which comes from NRC Emotion-Lexicon [14]. The depressed word dictionary is employed to classify social network text into two categories: those containing depressed words and those without depressed words. Each word in the text is assigned to one of these categories. The category assignment for each word is initialized randomly as a vector which we call sentiment marker bits: $S = (s_1, s_2, ..., s_n)$.

Word Embedding. Our word embedding is based on distributed representations of Words [13], mapping words into a high-dimensional feature space while preserving their semantic information. For each text, we represent it as $T = \{w_1, w_2, ..., w_n\}$ using word embedding, where $w_i \in \mathcal{R}^{\mathcal{D}}$ denotes each token embedding and $\mathcal{D}$ is dimensions of word vectors.

Due to the linear structure observed in typical word embedding representations, it becomes feasible to meaningfully combine words by element-wise addition of their vector representations. To effectively leverage the information contained within depressed words, we integrate each word embedding with the sentiment marker bits. Regarding the implementation aspect, a simple vector concatenation operation is used, and the embedding of a word v_i is calculated as $v_i = w_1 \bigoplus s_1$.

3.2 Multi-Task Learning Framework

Considering the diverse influences of various countries, regions, religions, and cultures, some meanings in many languages are often embedded within the underlying semantics rather than solely reflected in sentiment words. For instance, the word "blue" may not explicitly convey a sense of depression, but it often carries a pessimistic semantic meaning. As evident from the aforementioned example, depression text frequently includes negative sentiment words. However, relying solely on the sentiment information within the target sentence itself for depression detection often proves challenging in achieving satisfactory performance.

The task of determining the sentiment of a text based on its semantic information is commonly referred to as sentiment analysis. Extensive research has been conducted on sentiment analysis for many years, resulting in the availability of abundant high-quality labeled datasets. In contrast, in the depression detection field, the availability of high-quality labeled data is limited, leading to a restricted vocabulary and inherent biases during the training process. In multi-task learning, the commonly used framework employs a shared-bottom structure where different tasks share the bottom hidden layer. While this structure can mitigate overfitting risks, its effectiveness may be impacted by task dissimilarities and data distribution. In DeSK, we incorporate multiple identical feature extraction units that share the output from the previous layer as input and pass it on to the subsequent layer. This allows for an end-to-end training of the entire model. Our feature extraction units layer consists of a multi-head attention layer and two feed-forward neural networks.

Multi-head Attention Layer. To capture long-distance dependencies between words in a sentence, we employ the multi-head self-attention mechanism introduced by [25]. This approach calculates the semantic similarity and semantic features of each word in the sentence with respect to other words, allowing for enhanced connectivity and information exchange throughout the sentence. The formula is as follows:

For a given query $Q \in \mathcal{R}^{n_1 \times d_1}$, key $K \in \mathcal{R}^{n_1 \times d_1}$, value $V \in \mathcal{R}^{n_1 \times d_1}$,

$$Attention(Q, K, V) = softmax(\frac{QK^T}{d_1})V$$

For the i-th attention head, let the parameter matrix W_i^Q, W_i^K, W_i^V,

$$M_i = Attention(QW_I^Q, KW_I^K, VW_i^V)$$

The final feature representation is as follows:

$$H^s = concat(M_1, M_2, ..., M_l)W_o$$

Pooling Layer. Based on the observation [19] that using a combination of max-pooling and average-pooling yields significantly better performance compared to using a single pooling strategy alone. By leveraging both pooling strategies, we are able to capture different aspects and variations within the features, leading to improved overall performance. The formula is as follows:

$$P_m = Pooling_{max}(H^s)$$

$$P_a = Pooling_{average}(H^s)$$
$$P_s = concat(P_m, P_a)$$

3.3   Gated Attention Layer

The gated attention mechanism enables the model to dynamically select a subset of feature extraction units based on the input. Each task has its own gate, and the weight selection varies for different tasks. The output of a specific gate represents the probability of selecting a different feature extraction unit. Multiple units are then weighted and summed to obtain the final representation of the sentence, incorporating the contributions from the selected feature extraction units.

The formula is as follows:

$$g^k(x) = softmax(W_{gn} * gate(x))$$

$$f^k(x) = \sum_{i=1}^{n} g^k(x)_i f_i(x)$$

$$y_k = h^k f^k(x)$$

where k is the number of tasks.

3.4 Model Training

For the training process, the loss function used in DeSK combines cross entropy with L2 regularization, as follows:

$$loss = -\sum_i \sum_j y_i^j \log \hat{y}_i^j + \lambda \|\theta\|^2$$

where i is the index of sentences, and j is the index of class.

4   Experiment

In this section, we begin by introducing the datasets used in our study as well as the evaluation metrics employed for performance assessment. Then, we present a series of ablation experiments conducted to showcase the effectiveness of DeSK. Finally, we provide a comprehensive analysis of the results obtained from these experiments.

4.1 Datasets

To explore whether sharing sentiment knowledge can enhance the performance of hate depression detection, we utilize three publicly available datasets for social network depression detection and one sentiment dataset. Meanwhile, to validate the cross-language nature of the model, we use another depression detection dataset which we constructed, and a sentiment analysis dataset in Chinese. The specifics of these datasets are presented in Table 1.

Reddit Depression Dataset (RDD) is collected [18] from the archives of subreddit groups such as "r/MentalHealth," "r/depression," "r/loneliness," "r/stress," and "r/anxiety." These subreddits provide online platforms where individuals share their experiences and discussions related to mental health issues. The collected postings data were annotated by two domain experts who assigned three labels to denote the level of signs of depression: "Not depressed," "Moderate," and "Severe."

Tweet Depression Dataset 60k (TDD-60k) utilized in our study consists of four depression labels [22], each corresponding to different levels of depression signs. By incorporating these four depression labels, it aims to capture a comprehensive range of depression severity in the Twitter dataset.

Tweet Depression Dataset 10k (TDD-8k) is from huggingface[2]. The dataset exhibits a near-equivalent distribution of positive and negative examples, indicating a balanced representation between the two classes.

Sentiment Analysis (SA) is from Kaggle2018[3]. The SA dataset exhibits a higher number of positive cases but a relatively smaller number of negative cases. As the test set does not have labeled data, we solely rely on the training set for our analysis and model training.

Chinese Weibo Depression Dataset (CWDD) was compiled by collecting tweets from depression-related communities on Weibo. It underwent annotation and organization to ensure a balanced distribution of positive and negative cases, thereby achieving comparable proportions between the two classes. The dataset consists of 10,348 tweets from Weibo classified as depressed, while there are 7,562 tweets categorized as non-depressed.

Weibo Sentiment 100k (WS) consists of over 100,000 comments from Sina Weibo, which have been tagged with emotion labels. It contains 59,993 positive comments and 59,995 negative comments, making it a balanced dataset in terms of positive and negative sentiment. In consideration of the proportion of sentiment data and depression data, we have specifically chosen 10,500 instances from the dataset for the purpose of training. This selection process ensures a balanced representation of both sentiment and depression-related data in the training set.

Given the cross-language nature of our experiments, it is essential to construct separate depression word dictionaries for English and Chinese. The English depression dictionary is constructed with reference to the NRC Emotion Lexicon [14]. The NRC Emotion Lexicon is a widely recognized resource that provides a comprehensive collection of words annotated with their associated emotions. Our Chinese depression word dictionary is constructed based on the Dalian University of Technology Chinese Emotion Vocabulary Ontology Database [27]. This database is a recognized and comprehensive resource that contains a wide range of Chinese words and phrases annotated with their corresponding emotional categories.

4.2 Baselines and Metrics

Baselines. We compare the performance of DeSK with the following baselines to evaluate its effectiveness.

Doc2vec is proposed by [13], known as an unsupervised learning algorithm that aims to represent documents as fixed-length numerical vectors. It is an extension of the popular Word2Vec algorithm, which is used to generate word embeddings.

BERT is proposed by [6]. The pre-trained model BERT was used to capture the features of depression detection.

---

[2] https://huggingface.co/datasets/ShreyaR/DepressionDetection.
[3] https://www.kaggle.com/dv1453/twitter-sentiment-analysis-analytics-vidya.

Table 1. Statistics of datasets used in the experiment.

Dataset	total	Classes
RDD	156,676	Not depressed (4,649) Moderately depressed (10,494) Severely depressed (1,489)
TDD-60k	66,228	Not depressed (27,202) Moderately depressed(32,121) Severely depressed (9,968)
TDD-7k	7,771	Not Depressed (3,900) Depressed (3,832)
CWDD	17,910	Not Depressed (10,348) Depressed (7,562)
SA	31,962	negative (2,242) positive (29,720)
WS	119,988	negative (59,995) positive (59,993)

RoBERTa, proposed by [11], is an enhanced version of the BERT model, which introduces various optimizations to improve its performance. These optimizations focus on refining the underlying architecture and training process.

Metrics. In the depression detection task, we employ two evaluation metrics, namely Accuracy (ACC) and Macro F1, to assess the performance of DeSK.

4.3   Training Details

In the experiments, we use the following configurations for different components of the model. In the input layer, we initialize all word vectors using Glove Common Crawl Embeddings (840B Token) with a dimension of 300. The category embeddings, on the other hand, are initialized randomly with a dimension of 100.

For the sentiment knowledge sharing layer, we employ a multi-head attention mechanism with four heads. The first Feed-Forward network consists of a single layer with 400 neurons, while the second Feed-Forward network includes two layers with 200 neurons each. Dropout is applied after each layer with a dropout rate of 0.1.

The RMSprop optimizer is utilized with a learning rate of 0.001, and the models are trained using mini-batches consisting of 512 instances. To prevent overfitting, we incorporate learning rate decay and early stopping techniques during the training process. These measures ensure effective model training and help mitigate the risk of overfitting.

4.4 Model Performance

Table 2. Comparison of DeSK performance with the baseline in RDD.

Model	Accuracy	Macro F1
Doc2vec	54.1	54.9
BERT	55.9	56.1
RoBERTa	55.7	56.3
DeSK	**62.7**	**60.8**

The overall performance comparison is summarized in Table 2. DeSK outperforms other neural network models in terms of accuracy and F1 score. Specifically, compared to Doc2Vec, DeSK achieves a 10% increase in the F1 score. Even when compared to the strong baseline model, universal encoder, DeSK demonstrates superior performance. Furthermore, DeSK has the advantage of being easier to implement and having fewer parameters compared to other models. This makes it more accessible and efficient for practical applications.

Table 3. Performance of DeSK on different datasets.

Dataset	Accuracy	Macro F1
RDD	62.7	60.8
TDD-60k	81.9	80.9
TDD-7k	89.6	89.1
CWDD	96.3	96.3

Additionally, we conducted tests on different datasets, including the Chinese Weibo dataset, to evaluate the cross-language capability of DeSK. The results are shown in Table 3, indicating that DeSK performs well in analyzing and processing text from different languages, showcasing its ability to handle multilingual data effectively. This versatility further enhances the applicability of DeSK across various linguistic contexts.

4.5   Ablation Experiments

Table 4. The results of ablation experiments on TDD-60k dataset.

Model	Accuracy	Macro F1
-ss	77.4	77.3
-s	80.3	80.2
DeSK	**81.9**	**80.9**

We conducted an analysis to assess the impact of different components of DeSK. The results are presented in Table 4, where "-ss" refers to the ablation of sentiment knowledge sharing and sentiment marker bits, while "-s" indicates that sentiment data was not utilized as input and only sentiment marker bits were used in the model. The findings provide insights into the contributions of these components to the overall performance of the model.

Table 5. The influence of gate mechanism on TDD-7k dataset.

Model	Accuracy	Macro F1
-gate	89.2	88.8
DeSK	**89.6**	**89.1**

We conducted an analysis to evaluate the impact of gated attention in DeSK. The results, as shown in Table 5, demonstrate that the model's performance is further improved when gated attention is utilized. By learning how the outputs of different gates interact and contribute to the final representation, DeSK can better understand the dependencies and correlations between tasks, leading to improved performance in capturing complex task relationships.

5   Discussion

Depression has emerged as a significant global health concern, affecting individuals across various geographical locations. With the widespread adoption of social networks and the optional anonymity they provide, many individuals, whether diagnosed or not, may express their mood or symptoms related to depression on these platforms. This presents an opportunity to identify individuals at risk of depression through their online activities. Detecting individuals at risk of depression on social networks has great potential. It allows for timely intervention and support for those who may struggle to access social support or effective treatment through traditional means. By leveraging the power of technology and analyzing online behavior, we have the potential to reach individuals who might otherwise go unnoticed and provide them with the necessary assistance they need.

6   Conclusion and Future Work

This paper focuses on investigating the effectiveness of multi-task learning in depression detection tasks. The core concept revolves around utilizing multiple feature extraction units to share multi-task parameters, thereby enabling improved sharing of sentiment knowledge. The proposed model incorporates gated attention to fuse features for depression detection. By leveraging both the sentiment information from the target and external sentiment resources, DeSK demonstrates enhanced system performance through ablation experiments, thereby advancing social network depression detection. Through detailed analysis, we provide further evidence of the validity and interpretability of DeSK.

Overall, our experiments contribute to a better understanding of the interplay between depression detection and sentiment analysis through multi-task learning. They lay the foundation for future endeavors in refining modeling techniques and data selection, encompassing different types of social network depression information, diverse sentiment data types and scales, and other related aspects. But there are still some limitations. One key limitation is the quality of the available depression detection datasets, which are often labeled based on self-diagnosis. This introduces potential biases and uncertainties in the data, which can impact the performance of depression detection. Additionally, while psychologists may have access to a wealth of depression-related information through counseling sessions, privacy concerns prevent the correlation of this information with social network data. This limits our ability to create a comprehensive and high-quality depression detection dataset that combines both social network information and professional insights.

To overcome these limitations, we plan to explore privacy computing techniques to achieve alignment between social network data and individual entities without compromising privacy. This would allow us to create a more robust and reliable depression detection dataset, enhancing the accuracy and effectiveness of depression detection.

Acknowledgments. This work is supported by the National Key Research and Development Program of China (2021YFC3300500).

References

1. Chancellor, S., De Choudhury, M.: Methods in predictive techniques for mental health status on social media: a critical review. NPJ Digit. Med. **3**(1), 43 (2020)
2. Chandola, T., Kumari, M., Booker, C.L., Benzeval, M.: The mental health impact of covid-19 and lockdown-related stressors among adults in the UK. Psychol. Med. **52**(14), 2997–3006 (2022)
3. Cong, Q., Feng, Z., Li, F., Xiang, Y., Rao, G., Tao, C.: Xa-bilstm: a deep learning approach for depression detection in imbalanced data. In: 2018 IEEE International Conference on Bioinformatics and Biomedicine (BIBM), pp. 1624–1627. IEEE (2018)
4. Cuijpers, P., Stringaris, A., Wolpert, M.: Treatment outcomes for depression: challenges and opportunities. Lancet Psychiatry **7**(11), 925–927 (2020)

5. De Choudhury, M., Gamon, M., Counts, S., Horvitz, E.: Predicting depression via social media. In: Proceedings of the International AAAI Conference on Web and Social Media, vol. 7, pp. 128–137 (2013)
6. Devlin, J., Chang, M.W., Lee, K., Toutanova, K.: Bert: pre-training of deep bidirectional transformers for language understanding. arXiv preprint arXiv:1810.04805 (2018)
7. Gaind, B., Syal, V., Padgalwar, S.: Emotion detection and analysis on social media. arXiv preprint arXiv:1901.08458 (2019)
8. Guo, Z., Ding, N., Zhai, M., Zhang, Z., Li, Z.: Leveraging domain knowledge to improve depression detection on Chinese social media. IEEE Trans. Comput. Soc. Syst. (2023)
9. Lin, C., et al.: Sensemood: depression detection on social media. In: Proceedings of the 2020 International Conference on Multimedia Retrieval, pp. 407–411 (2020)
10. Liu, X., He, P., Chen, W., Gao, J.: Multi-task deep neural networks for natural language understanding. In: Proceedings of the 57th Annual Meeting of the Association for Computational Linguistics, pp. 4487–4496 (2019)
11. Liu, Y., et al.: Roberta: a robustly optimized bert pretraining approach. arXiv preprint arXiv:1907.11692 (2019)
12. Malviya, K., Roy, B., Saritha, S.: A transformers approach to detect depression in social media. In: 2021 International Conference on Artificial Intelligence and Smart Systems (ICAIS), pp. 718–723. IEEE (2021)
13. Mikolov, T., Sutskever, I., Chen, K., Corrado, G.S., Dean, J.: Distributed representations of words and phrases and their compositionality. Adv. Neural Inf. Process. Syst. **26** (2013)
14. Mohammad, S.M., Turney, P.D.: NRC emotion lexicon. Natl. Res. Council Canada **2**, 234 (2013)
15. Mouchet-Mages, S., Baylé, F.J.: Sadness as an integral part of depression. Dialog. Clin. Neurosci. (2022)
16. Orabi, A.H., Buddhitha, P., Orabi, M.H., Inkpen, D.: Deep learning for depression detection of twitter users. In: Proceedings of the Fifth Workshop on Computational Linguistics and Clinical Psychology: From Keyboard to Clinic, pp. 88–97 (2018)
17. Park, M., Cha, C., Cha, M.: Depressive moods of users portrayed in twitter. In: Proceedings of the 18th ACM International Conference on Knowledge Discovery and Data Mining (SIGKDD 2012), pp. 1–8 (2012)
18. Sampath, K., Durairaj, T.: Data set creation and empirical analysis for detecting signs of depression from social media postings. In: Kalinathan, L.R., Priyadharsini, K., Madheswari, S.M. (eds.) Computational Intelligence in Data Science: 5th IFIP TC 12 International Conference, ICCIDS 2022, Virtual Event, 24–26 March 2022, Revised Selected Papers, pp. 136–151. Springer, Cham (2022). https://doi.org/10.1007/978-3-031-16364-7_11
19. Shen, D., et al.: On the use of word embeddings alone to represent natural language sequences (2018)
20. Shen, G., et al.: Depression detection via harvesting social media: a multimodal dictionary learning solution. In: IJCAI, pp. 3838–3844 (2017)
21. Suhasini, M., Srinivasu, B.: Emotion detection framework for twitter data using supervised classifiers. In: Raju, K.S., Senkerik, R., Lanka, S.P., Rajagopal, V. (eds.) Data Engineering and Communication Technology. AISC, vol. 1079, pp. 565–576. Springer, Singapore (2020). https://doi.org/10.1007/978-981-15-1097-7_47
22. Tavchioski, I., Robnik-Šikonja, M., Pollak, S.: Detection of depression on social networks using transformers and ensembles. arXiv preprint arXiv:2305.05325 (2023)

23. Tølbøll, K.B.: Linguistic features in depression: a meta-analysis. J. Lang. Works Sprogvidenskabeligt Studentertidsskrift **4**(2), 39–59 (2019)
24. Tsugawa, S., Kikuchi, Y., Kishino, F., Nakajima, K., Itoh, Y., Ohsaki, H.: Recognizing depression from twitter activity. In: Proceedings of the 33rd Annual ACM Conference on Human Factors in Computing Systems, pp. 3187–3196 (2015)
25. Vaswani, A., et al.: Attention is all you need. Adv. Neural Inf. Process. Syst. **30** (2017)
26. Wongkoblap, A., Vadillo, M.A., Curcin, V.: Modeling depression symptoms from social network data through multiple instance learning. AMIA Summit. Transl. Sci. Proc. **2019**, 44 (2019)
27. Xu, L., Lin, H., Pan, Y., Ren, H., Chen, J.: Constructing the affective lexicon ontology. J. China Soc. Sci. Tech. Inf. **27**(2), 180–185 (2008)
28. Ypsilanti, A., Lazuras, L., Powell, P., Overton, P.: Self-disgust as a potential mechanism explaining the association between loneliness and depression. J. Affect. Disord. **243**, 108–115 (2019)
29. Yue, L., Chen, W., Li, X., Zuo, W., Yin, M.: A survey of sentiment analysis in social media. Knowl. Inf. Syst. **60**, 617–663 (2019)
30. Zhang, Y., Yang, Q.: A survey on multi-task learning. IEEE Trans. Knowl. Data Eng. **34**(12), 5586–5609 (2021)
31. Zhang, Z., Luo, P., Loy, C.C., Tang, X.: Facial landmark detection by deep multi-task learning. In: Fleet, D., Pajdla, T., Schiele, B., Tuytelaars, T. (eds.) ECCV 2014. LNCS, vol. 8694, pp. 94–108. Springer, Cham (2014). https://doi.org/10.1007/978-3-319-10599-4_7
32. Zhou, X., et al.: Hate speech detection based on sentiment knowledge sharing. In: Proceedings of the 59th Annual Meeting of the Association for Computational Linguistics and the 11th International Joint Conference on Natural Language Processing, vol. 1: Long Papers, pp. 7158–7166 (2021)

AOM: A New Task for Agitative Opinion Mining in We-media

Huazi Yin⬤, Jintao Tang$^{(\boxtimes)}$⬤, Shasha Li⬤, and Ting Wang⬤

College of Computer Science and Technology, National University of Defense
Technology, Changsha 410073, China
{huaziyin,tangjintao,shashali,tingwang}@nudt.edu.cn

Abstract. As We-media continues to develop, there is a concerning rise
of agitative opinions on We-media platforms, which usually lead to online
violence. To formalize the research on identifying whether a sentence contains agitative opinions, we propose a new task AOM for agitative opinion mining in We-media. To clarify the task, we make a clear definition
to agitative opinions in which they are categorized into nine types and
we manually construct a ten-thousands scale Chinese agitative opinion
dataset CAOD based on WeChat public account, for research purpose.
Furthermore, a baseline model CAOD$_{\mathrm{MINER}}$ based on TextCNN is proposed and sampling methods are adopted in training it. For comparison,
we also apply several mainstream text classifiers into CAOD. The comparative experiment and further analysis show that AOM is a soluble
but challenging task, where unbalanced data distribution, diversity of
expression forms, context dependency, scarcity of external knowledge
and implicit expression deserve to be studied in the future.

Keywords: Agitative opinion mining · We-media · Chinese agitative
opinion dataset

1 Introduction

In recent years We-media has become the main channel of information dissemination. People are increasingly willing to express their opinions on We-media
platforms. However, due to the informal and anonymous feature of We-media,
opinions that cause bad effect to receivers are also easily expressed and disseminated. Particularly, in order to get more attention, some We-media publish
articles which involve a large number of agitative opinions to increase hits. Agitative opinion is typically the speech or language that causes or tends to cause
crowds' anger or anxiety and forces them to follow. For example, some We-
media accounts deliberately exaggerate the news about vaccine side effects and
post agitative opinions such as "vaccines are the biggest conspiracy in history",
to attract the attention of concerned parents who want to vaccinate their children. Especially towards the hot events, some We-medias guide netizens to vent
negative emotions upon the innocent by exaggerating and creating antagonism,

© The Author(s), under exclusive license to Springer Nature Singapore Pte Ltd. 2024
F. Wu et al. (Eds.): SMP 2023, CCIS 1945, pp. 147–160, 2024.
https://doi.org/10.1007/978-981-99-7596-9_11

which usually causes online violence. Therefore, it is necessary to detect the agitative opinions in We-media. The automatic identification of agitative opinions could not only provide reading help for netizens to prevent them from being misled, but also be applied to the review of articles on social media platforms to improve the overall quality of We-media articles and create a clear network space.

Opinion Mining, which is a popular research field in Natural Language Processing, is defined by Liu and Zhang [20] as "the computational study of people's opinions, appraisals, attitudes, and emotions toward entities, individuals, issues, events, topics and their attributes". Currently, there are four mainstream tasks in opinion mining field which are sentiment classification, feature-based opinion mining, comparative relationship mining and opinion retrieval. Among these four tasks, the later three tasks [15,16] mainly aim to mine people's opinions on certain things to evaluate how good or bad they are, which are strongly domain-oriented. By contrast, the sentiment classification task which infers the sentimental polarity of a message or review [2,25] is the most relevant to agitative opinion mining. For example, Lyv et al. [21] employed LSTM, BERT and ERNIE methods to predict the sentiments of Weibo users responding to COVID-19, including fear, anger, disgust, sadness, gratitude, surprise, and optimism. However, sentiment classification task focuses on how people feel such as anxiety, disgust and anger, with less consideration about detecting the opinions causing these feelings like hate speech, rumor and other agitative opinions.

In recent years, some researches relevant to detecting agitative opinions have been proposed. Offensive language and hate speech are highly related agitative opinions with blurred boundaries [12], which both denigrate or offend an individual or group, using expressions of racism, sexism, personal attacks, harassment, and cyberbullying [29]. Hajibabaee et al. [9] proposed a model consisting of eight classifiers to detecting offensive language. Deng et al. [5] constructed a benchmark COLD for Chinese offensive language detection. Rumor, which is defined as a statement whose truth value is unverified or deliberately false [27], is another kind of agitative opinions that causes bad effect to receivers. To detect rumors on source post, Miao et al. [24] proposed Syntax and Sentiment Enhanced BERT (SSE-BERT). Although these researches aim to detect the bad opinions that are agitative and may cause violence, anxiety and blind conformity, there is still no clear answer to what agitative opinions are and how to detect them automatically.

To fill in the blanks in researches on detecting agitative opinions, in this paper we present a new task AOM for agitative opinion mining in We-media. Through survey and summarisation, we categorize the agitative opinions into nine types including offensive language, rumor and other opinions, and we construct a ten-thousands scale Chinese agitative opinion dataset CAOD based on a widely-used We-media platform WeChat public account. Furthermore, we propose a baseline model $CAOD_{MINER}$ to identify whether a sentence contains agitative opinions based on TextCNN. Over-sampling and under-sampling methods are also adopted to address the unbalanced data distribution problem of online agitative opinions. In comparative experiments, we migrate several mainstream models widely used in text classification to CAOD. The experiment results indicate that AOM is a complicated task and there are many challenges deserved to be further studied. The contributions of our work are as follows:

- We propose AOM task for agitative opinion mining in We-media and give clear definition of agitative opinions.
- We construct a Chinese agitative opinion dataset CAOD [1] based on WeChat public account, for research purpose.
- We propose a baseline model CAOD$_{\text{MINER}}$ and conduct comparative experiments, confirming that AOM is a soluble but challenging problem, where unbalanced data distribution, diversity of expression forms, context dependency, scarcity of external knowledge and implicit expression deserve to be further studied.

2   Related Work

2.1   Opinion Mining

Sentiment classification, which is a popular problem in opinion mining field, has been studied for many years. Early approaches of sentiment classification are lexicon-based approaches and traditional machine learning algorithms. Taboada et al. [31] adopted dictionaries of words annotated with semantic orientation for sentiment classification. Huang et al. [11] used Support Vector Machine (SVM) to analyze sentiment of user-generated text in financial field. Xia et al. [33] applied a Multi-Layer Perceptron to classify tweets about the 2020 U.S. Presidential Election with a positive or negative sentiment. Recently, deep neural network is gradually used. Hao et al. [10] presented a Word2vec and Dynamic Conditional Random Field (DCRF) based framework for subjectivity classification and polarity classification of Chinese Microblog. Feature-based opinion mining is more fine-grained which extracts aspects and analyzes corresponding sentiment. For example, Pang et al. [26] proposed a novel Highway-Based Local Graph Convolution Network (HL-GCN) to identify the sentiment polarity of a specific aspect in a sentence. Comparative relationship mining and opinion retrieval are often conducted using rule-based and machine learning approaches. Feldman et al. [8] extracted information about product comparisons using extracting patterns. Liao et al. [19] used statistical machine learning methods to retrieve documents that were relevant to the query topic. However, these four tasks of opinion mining mainly focus on analysis of human's emotions (positive/negative or happy/sad/angry) and their evaluations of how good or bad things are in various aspects. The detection of agitative opinions which cause people's negative feelings is not well studied.

---

[1] https://github.com/hzyinn/Chinese-agitative-opinion-dataset.

2.2   Offensive Language and Rumor Detection

As two typical agitative opinions with bad effect online, detection of offensive language and rumor are gradually studied. Automatic approaches for offensive language detection can be divided into key-words-based approach and machine learning classifier. Key-words-based approach identifies the hateful keywords using an ontology or dictionary, such as Hatebase [1]. Machine learning classifier is a machine learning model trained on annotated hateful texts. For example, Davidsm et al. [4] proposed a feature-based classification model using SVM. Pre-trained models like BERT [22] and C-GRU [35] have also been used in offensive language detection. Corresponding to rumor's life-cycle, detection of rumor can be aligned with three stages that are earlist, early and general stage. Most researches focus on the latter two stages, combining source posts with reply posts to detect rumor. Khoo et al. [13] detected rumor by extracting interaction features from source and reply posts using a structure aware self-attention model. Wang et al. [32] proposed SD-TsDTS-CGRU model for rumor events detection, with people's emotional reactions considered. Few researches study the earliest stage where only the source post can be used [24]. Although increasing researches on offensive language and rumor detection have been proposed, there are still more types of agitative opinions deserved to be explored and automatically detected. Accordingly, we propose a new task named as AOM to formalize agitative opinion mining, and we discuss the challenges of this task which can be further researched.

3   Task Definition

3.1   Agitative Opinion Types

According to researches [7,17,23,30] in journalism and communication field, we categorize agitative opinions into nine types shown in Table 1. Opinions of any type in Table 1 are regarded as agitative opinions, which gives a standard for data annotation. For instance, if an opinion of the writer tends to influence the readers' mind and force them to do something, we regard it as an inflammatory opinion, which is also agitative. The sentence "If you don't forward this article, you are not Chinese" forces people to forward the article even if useless, so it has inflammatory and agitative opinions. Taking another sentence from a We-media article as example, "In fact, we are in a 'thin is beautiful' era." This sentence not only causes the readers' body anxiety, suggesting that overweight people are not beautiful, but also overgenerically believes that this era is totally thin for beauty, conveying unrealistic information. The fearmongering and unrealistic opinions in this sentence are both agitative opinions defined in Table 1, so this sentence is considered to contain agitative opinions. For the other types of agitative opinions such as group-opposition, offensive and so on, we also give clear definition in Table 1 and provide several examples to better explain.

3.2 Problem Formulation

Let $D = \{(s_1, y_1), (s_2, y_2), \ldots, (s_n, y_n)\}$ be a dataset, where s_i denotes the i-th sentence that may contain agitative opinions or not, y_i is the corresponding label and n denotes the size of dataset. We define AOM task as a sentence-level binary classification task viz. $y_i \in \{0, 1\}$, where $y_i = 1$ represents s_i contains agitative opinions defined in Table 1, and $y_i = 0$ represents s_i not contains. Accordingly, AOM can be described as: given a sentence set $S = \{s_1, s_2, \ldots, s_n\}$, the corresponding label set $Y = \{y_1, y_2, \ldots, y_n\}$ needs to be predicted.

Table 1. Agitative opinion types.

Type	Definition	Example
Inflammatory	To influence someone's mind and force them to act as expected	You must; Believe in xx and have eternal life, etc.
Fearmongering	To make people anxious by something, conveying anxiety, frustration and self-deprecation emotions	Starting fare of love; Poverty limits imagination, etc.
Group opposition	To use conflict and opposition between groups or ranks to attract attention and earn traffic [17]	Rich second generation and the Short, poor and ugly; loser; bottom people, etc.
Unrealistic	To deliberately distorting and fabricating facts in order to attract attention, or retweeting hearsay without verifying it [23]	It is understood that the development process of rural areas is completely out of step with the times, etc.
Offensive	To denigrate or offend an individual or group, using expressions of racism, sexism, personal attacks, harassment, and cyberbullying [30]	Brute; Niubi; TMD, etc.
Pornographic	To Talk about sex and pornography, wondering on the edge of law and morality [17]	Someone puts his hand down my pants, etc.
Mammonish	To regard money as the highest principle of life [7]	Rich is great; Stop touching it since you cannot afford it, etc.
Immoral	To ignore social duty, distort correct values and challenge the bottom line of social morality in order to attract attention	Visiting prostitutes is a way of socializing, etc.
Extreme	To treat complex social phenomena as a simple story or conclusion, deliberately creating ideological binary opposition [17]	Couples are exactly at their best when they love based on money, etc.

4　Dataset Construction

To obtain plenty of We-media articles that may contain agitative opinions for dataset construction, we crawl WeChat public account, which is a widely-used We-media platform. We use keywords of recent hot events such as "northern Myanmar" and "actor tax evasion" as indexes to obtain recent We-media articles. As the agitative opinions defined in Table 1 mostly contain negative sentiments, we first select the negative articles to ensure that enough sentences containing agitative opinions can be extracted from these articles. Since there has been much work using BERT [6] as foundational model for sentiment classification and getting good performance [3,34], we adopt BERT combined with bi-directional long-short term memory network in selecting the negative articles. To avoid error accumulation, auxiliary manual inspection is also applied to selection.

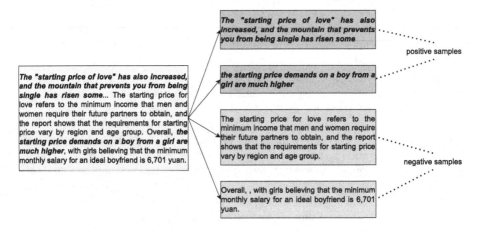

Fig. 1. Data annotating and extracting process.

After selecting 997 negative articles, the sentences containing agitative opinions are annotated manually according to the classification in Table 1. There are two Chinese natives involved in annotation and when their annotations differ, they discuss to determine the final label. Finally, the sentences labeled as containing agitative opinions are extracted to form the positive samples with the rest negative ones. Figure 1 shows the process of annotating and extracting. The first sentence in the middle column conveys anxiety to singles that they may be not rich enough to start a relationship. The second sentence draws the conclusion that girls require much higher starting price than boys merely from one report, which is unrealistic and unconvincing. And it causes group-opposition between men and women to a certain extent. These two sentences both contain agitative opinions, so they are annotated and extracted directly from the left paragraph. The rest paragraph are then divided into sentences with no agitative opinions to form the negative samples.

Table 2. Statistics of CAOD.

	Positive samples	Negative samples	Whole dataset
Statistics	2198 (8.3%)	24394 (91.7%)	26592

The statistics of our Chinese agitative opinion dataset CAOD are shown in Table 2. There are totally 26,592 samples in CAOD, with 2,198 positive samples and 24,394 negative samples.

5 Agitative Opinion Mining

5.1 Baseline Model

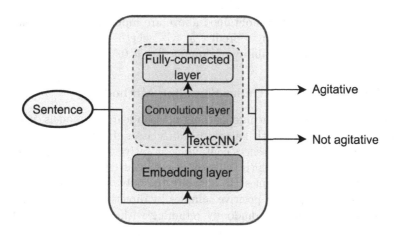

Fig. 2. Framework of CAOD$_{\text{MINER}}$.

We propose a baseline model CAOD$_{\text{MINER}}$ for CAOD based on TextCNN, whose structure is shown in Fig. 2. For an input sentence s_i, pre-trained Chinese word embedding [18,28] is used to encode the sentence, forming a vector matrix X_i. Since a sentence containing agitative opinions usually includes several agitative keywords, we choose convolutional neural network which can focus on different local features of the input to classify whether a sentence contains agitative opinions. Therefore, the obtained vector matrix X_i is input into TextCNN [14] to classify whether the sentence contains agitative opinions. TextCNN is mainly composed of multiple groups of multi-channel convolutional layers, pooling operation, dropout operation and fully connected layer.

Firstly, three groups of convolution kernels with height two, three, and four are set, so that a window contains two, three, and four words respectively, corresponding to the 2-gram, 3-gram, and 4-gram models of the text. Secondly, in order to learn the features of different aspects of the text, $num\_filters$ channels are set up in each group of convolution kernels, where $num\_filters = 256$. For the convolution kernel of each channel in each group, its height is denoted as h, $h \in 2, 3, 4$. Formula 1 and Formula 2 describe the convolution result of the convolution kernel for the sequence X_i:

$$c_{i_j} = ReLU(w \cdot X_{i_{j:j+h-1}} + b) \tag{1}$$

$$c_i = c_{i_1}, c_{i_2}, \dots, c_{i_{m-h+1}} \tag{2}$$

By performing the above operation on X_i, $3 * num\_filters$ convolution results can be obtained. In order to preserve the most obvious features of the text, max pooling operations are performed on the $3 * num\_filters$ convolution results obtained above, as shown in Formula 3.

$$c_{i_{max}} = max\{c_i\} \tag{3}$$

Afterwards, the feature values obtained above are input into a fully connected layer after dropout operation to predict y_i, viz. whether the sentence contains agitative opinions. The probability of dropout is $p = 0.5$.

5.2   Over-Sampling and Under-Sampling Methods

In the process of training machine learning or deep learning models, it is often preferred to use datasets with balanced data classes. However, on real online We-media platforms, the sentences containing agitative opinions are quite fewer than those not containing, which leads to a significant difference in the number of positive and negative samples. To alleviate the unbalanced data distribution problem, we over-sample the positive samples by directly repeating and under-sample the negative ones by randomly retaining with a probability of 50% in training set and retrain CAOD$_{\text{MINER}}$.

5.3   Experiment Setup

Dataset. Experiments are conducted on CAOD constructed in Sect. 4. The size of the split training set is 23,933 and the size of the testing set is 2,659.

Comparative Models. To discuss how mainstream text classification models perform on CAOD, we train and test several mainstream models on CAOD. 1. SVM: support vector machine. 2. fastText: fast text classification model based on shallow neural network. 3. BERT+Bi-LSTM: a model using BERT as encoder and finetuning in downstream tasks applying bi-directional long-short term memory network.

Evaluation Metrics. In this paper, four metrics including accuracy, precision, recall and F1 value are used to evaluate the effectiveness of models.

Table 3. Results of comparative experiments.

	Accuracy (%)	Precision (%)	Recall (%)	F1 (%)
SVM	92.93	20.45	**77.59**	32.37
fastText	93.16	33.64	67.27	44.85
BERT+Bi-LSTM	94.05	71.94	45.66	55.87
CAOD$_{miner}$- sampling	**95.94**	**88.36**	58.64	70.49
CAOD$_{miner}$	95.83	82.63	62.73	**71.32**

5.4   Experiment Results

The results of comparative experiments are shown in Table 3. We can notice that CAOD$_{MINER}$ without sampling methods outperforms other models in accuracy, precision and F1 value. It proves that due to the use of TextCNN, CAOD$_{MINER}$ can learn the local sequence features of the agitative keywords efficiently. Furthermore, to alleviate the unbalanced data distribution of We-media platforms, we adopt over-sampling and under-sampling methods, further improving the recall by 4.09% and the F1 value by 0.83%, which confirms that AOM is a soluble problem.

Table 3 also indicates that the mainstream text classifiers cannot be easily transferred to AOM with good performance. Although they achieve slightly more than 90% accuracy rate, their F1 values are less than 60% on CAOD. The reason for the low F1 value is that the unbalanced data distribution of about 1:11 between positive and negative samples makes the model learn the features of the negative samples comparatively better than the positive ones, leading to either a low precision or a low recall.

5.5   Error Analysis

To further discuss the challenges of AOM, we qualitatively analyze the mistaken samples of CAOD$_{MINER}$ in the test set of CAOD and summarize the following challenges of AOM deserved to be further studied.

Table 4. Results of positive samples and negative samples.

	Number	Precision (%)	Recall (%)	F1 (%)
Positive samples	220	82.63	62.73	71.32
Negative samples	2440	96.71	98.81	97.75

Unbalanced Data Distribution. Table 4 presents predicting results of positive samples and negative samples respectively, which indicates that $CAOD_{MINER}$ identifies negative samples better. The reason is that on the real online We-media platforms like WeChat public account, the sentences containing agitative opinions are much fewer than those not containing. The unbalanced data distribution makes models difficult to learn the features of agitative opinions. Although the ratio is adjusted to 1:2.8 from 1:11 in our over-sampling and under-sampling methods, this problem still exists. $CAOD_{MINER}$ learns the features of negative samples more adequately, while it does not learn the features of positive examples, which account for a smaller proportion, very well.

Diversity of Expression Forms. There are 21.95% false negative(FN) samples misclassified due to the diversity of expressions with the same meaning. Internet phrases are informal and personal, with authors using different forms of expression to convey the same meaning. The model cannot learn all possible forms of expression adequately, so it is easy to get confused when identifying them. For example, some internet terms, such as insulting words, do not use the Chinese characters such as " 傻逼 " in order to pass scrutiny, but instead use the acronyms "SB", which convey the same offensive meaning. Since these vocabularies are embedded on word or letter level in the embedding layer of $CAOD_{MINER}$, and there are few positive samples containing the expression form of acronyms in the training process, the model is hard to understand these offensive acronyms and can easily confuse them with formal name abbreviations, making them difficult to identify.

Context Dependency. About 9.76% FN samples and 37.93% FP samples are mistaken because of scarcity of context dependency. The complexity and variability of Chinese lead to the fact that the same words or sentences may express different meanings in different contexts. There are also sentences that do not have an obvious emotional colors, but only contain agitative opinions when they occur in a specific context. The identification of these sentences is context dependent. Since context is taken into account in data annotation process but not when input into $CAOD_{MINER}$, the model may make mistakes when predicting whether a sentence contains agitative opinions. For example, the sentence "Their ultimate solution is always to give out red packets." itself unlikely contains agitative opinions. However, in the context of "It's unlikely that you'll be able to report unequal treatment in a company. Their ultimate solution is always to give out red packets. You're too embarrassed to write a letter after you've received a red packet, right?", this sentence contains agitative opinions of information distortion and overgeneralization. Without considering the context of this sentence, the model may make wrong judgments.

Scarcity of External Knowledge. Whether a sentence contains agitative opinions or not sometimes needs to be judged on the basis of prior external knowledge, which cannot be merely determined by models or context. 39.02% FN samples and 44.83% FP samples in the test set are wrongly identified owing to this reason. For example, "How can they possibly have a sense of responsibility, mission and honour for their country and team when losing a game is far more profitable than winning one?" The statement that "losing a game is far more profitable than winning one" requires external knowledge to determine whether it is a misinformed and agitative opinion.

Implicit Expression. Agitative opinions in 19.51% of FN samples are expressed implicitly without specific characteristics, making it difficult for the model to learn. For instance, some anxiety-producing statements do not directly express anxiety, but rather provoke the readers' anxiety by side-tracking or innuendo. For example, "If you insist on wearing super-shorts, I hope you can meet the following conditions - 1. having long and beautiful legs and 2. handsome" does not directly mean that "super-shorts are not for everyone". Instead, the readers are implied that not everyone can meet the requirements, with body anxiety triggered.

6 Conclusion

To protect netizens from agitative opinions' misguidance, we propose a new task named as AOM for agitative opinion mining in We-media. To clarify the task, we build a ten-thousands scale Chinese agitative opinion dataset CAOD by manual, in which the agitative opinions are categorized into nine types. Furthermore, we propose a baseline model CAOD$_{MINER}$ based on TextCNN and sampling methods. We also apply several mainstream text classifiers in CAOD for comparison. The results show that agitative opinions could be identified by technical approaches but there are still many challenges in AOM, such as unbalanced data distribution, variability and complexity of Chinese language and scarcity of external knowledge which deserve to be studied in the future. Moreover, since we obtain the sentences containing agitative opinions merely from negative articles to ensure that enough positive samples could be extracted, we will expand our dataset to include other sentences which contain agitative opinions extracted from positive articles. We will also explore the features of the nine types of agitative opinions and apply them to construct a more fine-grained classification model, further refining the AOM task and achieving automatic identification of what type of agitative opinions a sentence contains.

References

1. Hatebase, available from: https://hatebase.org/
2. Chen, L.S., Liu, C.H., Chiu, H.J.: A neural network based approach for sentiment classification in the blogosphere. J. Informet. **5**(2), 313–322 (2011)

3. Chen, N., Xia, Q., Zhou, X., Chen, W., Zhang, M.: Emotion classification with explicit and implicit syntactic information. In: Wang, L., Feng, Y., Hong, Y., He, R. (eds.) Natural Language Processing and Chinese Computing, pp. 607–618. Springer International Publishing, Cham (2021)

4. Davidson, T., Warmsley, D., Macy, M., Weber, I.: Automated hate speech detection and the problem of offensive language. In: Proceedings of the international AAAI conference on web and social media. vol. 11, pp. 512–515 (2017)

5. Deng, J., Zhou, J., Sun, H., Zheng, C., Mi, F., Meng, H., Huang, M.: Cold: A benchmark for chinese offensive language detection. arXiv preprint arXiv:2201.06025 (2022)

6. Devlin, J., Chang, M.W., Lee, K., Toutanova, K.: BERT: Pre-training of deep bidirectional transformers for language understanding. In: Proceedings of the 2019 Conference of the North American Chapter of the Association for Computational Linguistics: Human Language Technologies, Volume 1 (Long and Short Papers). pp. 4171–4186. Association for Computational Linguistics, Minneapolis, Minnesota (Jun 2019). 10.18653/v1/N19-1423, https://aclanthology.org/N19-1423

7. Fan, R.: Money worship and its spread in china. Theoretical Investigation **6**, 35–37 (1996)

8. Feldman, R., Fresko, M., Goldenberg, J., Netzer, O., Ungar, L.: Extracting product comparisons from discussion boards. In: Seventh IEEE international conference on data mining (ICDM 2007). pp. 469–474. IEEE (2007)

9. Hajibabaee, P., Malekzadeh, M., Ahmadi, M., Heidari, M., Esmaeilzadeh, A., Abdolazimi, R., James Jr, H.: Offensive language detection on social media based on text classification. In: 2022 IEEE 12th Annual Computing and Communication Workshop and Conference (CCWC). pp. 0092–0098. IEEE (2022)

10. Hao, Z., Cai, R., Yang, Y., Wen, W., Liang, L.: A dynamic conditional random field based framework for sentence-level sentiment analysis of chinese microblog. In: 2017 IEEE International Conference on Computational Science and Engineering (CSE) and IEEE International Conference on Embedded and Ubiquitous Computing (EUC). vol. 1, pp. 135–142 (2017). DOI: 10.1109/CSE-EUC.2017.33

11. Huang, X., Zhang, L.: An svm ensemble approach combining spectral, structural, and semantic features for the classification of high-resolution remotely sensed imagery. IEEE Trans. Geosci. Remote Sens. **51**(1), 257–272 (2012)

12. Jahan, M.S., Oussalah, M.: A systematic review of hate speech automatic detection using natural language processing. Neurocomputing p. 126232 (2023)

13. Khoo, L.M.S., Chieu, H.L., Qian, Z., Jiang, J.: Interpretable rumor detection in microblogs by attending to user interactions. In: Proceedings of the AAAI conference on artificial intelligence. vol. 34, pp. 8783–8790 (2020)

14. Kim, Y.: Convolutional neural networks for sentence classification. In: Proceedings of the 2014 Conference on Empirical Methods in Natural Language Processing (EMNLP). pp. 1746–1751. Association for Computational Linguistics, Doha, Qatar (Oct 2014). https://doi.org/10.3115/v1/D14-1181, https://aclanthology.org/D14-1181

15. Kumar, R., Pannu, H.S., Malhi, A.K.: Aspect-based sentiment analysis using deep networks and stochastic optimization. Neural Comput. Appl. **32**, 3221–3235 (2020)

16. Laddha, A., Mukherjee, A.: Extracting aspect specific opinion expressions. In: Proceedings of the 2016 Conference on Empirical Methods in Natural Language Processing. pp. 627–637 (2016)

17. LI, L.: The form, harm and governance of the implicit internet anomie behavior of self-media:a case analysis of "mi meng". Humanities & Social Sciences Journal of Hainan University 37(3), 73–79 (2019)

18. Li, S., Zhao, Z., Hu, R., Li, W., Liu, T., Du, X.: Analogical reasoning on Chinese morphological and semantic relations. In: Proceedings of the 56th Annual Meeting of the Association for Computational Linguistics (Volume 2: Short Papers). pp. 138–143. Association for Computational Linguistics, Melbourne, Australia (Jul 2018). 10.18653/v1/P18-2023 , https://aclanthology.org/P18-2023
19. Liao, X., Liu, D., Gui, L., Cheng, X., Chen, G.: Opinion retrieval method combining text conceptualization and network embedding. Journal of Software **29**(10), 2899–2914 (2018)
20. Liu, B., Zhang, L.: A survey of opinion mining and sentiment analysis. In: Mining text data, pp. 415–463. Springer (2012)
21. Lyu, X., Chen, Z., Wu, D., Wang, W.: Sentiment analysis on chinese weibo regarding covid-19. In: Natural Language Processing and Chinese Computing: 9th CCF International Conference, NLPCC 2020, Zhengzhou, China, October 14–18, 2020, Proceedings, Part I 9. pp. 710–721. Springer (2020)
22. MacAvaney, S., Yao, H.R., Yang, E., Russell, K., Goharian, N., Frieder, O.: Hate speech detection: Challenges and solutions. PLoS ONE **14**(8), e0221152 (2019)
23. Miao, T.: Ethical and moral review of the anomy of information communication in the we media era. Journalism Lover **5**, 61–63 (2016)
24. Miao, X., Rao, D., Jiang, Z.: Syntax and sentiment enhanced bert for earliest rumor detection. In: Natural Language Processing and Chinese Computing: 10th CCF International Conference, NLPCC 2021, Qingdao, China, October 13–17, 2021, Proceedings, Part I 10. pp. 570–582. Springer (2021)
25. Pang, B., Lee, L., Vaithyanathan, S.: Thumbs up? sentiment classification using machine learning techniques. arXiv preprint cs/0205070 (2002)
26. Pang, S., Yan, Z., Huang, W., Tang, B., Dai, A., Xue, Y.: Highway-based local graph convolution network for aspect based sentiment analysis. In: Natural Language Processing and Chinese Computing: 10th CCF International Conference, NLPCC 2021, Qingdao, China, October 13–17, 2021, Proceedings, Part I 10. pp. 544–556. Springer (2021)
27. Qazvinian, V., Rosengren, E., Radev, D., Mei, Q.: Rumor has it: Identifying misinformation in microblogs. In: Proceedings of the 2011 conference on empirical methods in natural language processing. pp. 1589–1599 (2011)
28. Qiu, Y., Li, H., Li, S., Jiang, Y., Hu, R., Yang, L.: Revisiting correlations between intrinsic and extrinsic evaluations of word embeddings. In: Chinese Computational Linguistics and Natural Language Processing Based on Naturally Annotated Big Data: 17th China National Conference. CCL 2018, and 6th International Symposium, NLP-NABD 2018, pp. 1–12. Changsha, China (2018)
29. Rajamanickam, S., Mishra, P., Yannakoudakis, H., Shutova, E.: Joint modelling of emotion and abusive language detection. arXiv preprint arXiv:2005.14028 (2020)
30. Rajamanickam, S., Mishra, P., Yannakoudakis, H., Shutova, E.: Joint modelling of emotion and abusive language detection. In: Proceedings of the 58th Annual Meeting of the Association for Computational Linguistics. pp. 4270–4279. Association for Computational Linguistics, Online (Jul 2020). 10.18653/v1/2020.acl-main.394 , https://aclanthology.org/2020.acl-main.394
31. Taboada, M., Brooke, J., Tofiloski, M., Voll, K., Stede, M.: Lexicon-Based Methods for Sentiment Analysis. Computational Linguistics **37**(2), 267–307 (06 2011). DOI: https://doi.org/10.1162/COLI_a_00049, https://doi.org/10.1162/COLI_a_00049
32. Wang, Z., Guo, Y.: Rumor events detection enhanced by encoding sentimental information into time series division and word representations. Neurocomputing **397**, 224–243 (2020)

33. Xia, E., Yue, H., Liu, H.: Tweet sentiment analysis of the 2020 us presidential election. In: Companion proceedings of the web conference 2021. pp. 367–371 (2021)

34. Zhang, S., Bai, X., Jiang, L., Peng, H.: Dual adversarial network based on bert for cross-domain sentiment classification. In: Wang, L., Feng, Y., Hong, Y., He, R. (eds.) Natural Language Processing and Chinese Computing, pp. 557–569. Springer International Publishing, Cham (2021)

35. Zhang, Z., Robinson, D., Tepper, J.: Detecting hate speech on twitter using a convolution-gru based deep neural network. In: The Semantic Web: 15th International Conference, ESWC 2018, Heraklion, Crete, Greece, June 3–7, 2018, Proceedings 15. pp. 745–760. Springer (2018)

PNPT: Prototypical Network with Prompt Template for Few-Shot Relation Extraction

Liping Li[1], Yexuan Zhang[1], Jiajun Zou[1(✉)], and Yongfeng Huang[1,2]

[1] Tsinghua University, Beijing 100084, China
{llp21,zhangyex21,zjj21}@mails.tsinghua.edu.cn, yfhuang@tsinghua.edu.cn
[2] Zhongguancun Laboratory, Beijing 100094, China

Abstract. Few-shot relation extraction involves predicting the relations between entity pairs in a sentence with a limited number of labeled instances for each specific relation. Prototypical network, which is based on the meta-learning framework, has been widely adopted for this task. Existing prototypical network-based approaches typically obtain the relation representation by concatenating the embeddings corresponding to start tokens of two entity mentions. While these methodologies have demonstrated commendable performance, we argue that the current relation representation fails to fully capture semantic nuances within complex scenes, where the identical entity pairs often convey diverse semantic relationship. In this paper, we propose an innovative relation representation approach that integrates textual context and entity mentions through a prompt template. Furthermore, we introduce a gate mechanism to selectively incorporate external relation knowledge into the origin relation prototype derived from support instances. Experimental results on two benchmark datasets demonstrate the effectiveness of our proposed approach.

Keywords: Relation Extraction · Few Shot · Prompt Template

1 Introduction

Relation extraction (RE) [12] is a challenging task in Natural Language Processing (NLP), which aims to classify the relationship between entity pairs contained in sentences and can be applied to various downstream applications, including knowledge graph construction [9] and question answering [2]. However, RE usually contends with the challenge of training data scarcity due to the time-consuming and labor-intensive of data annotation. To overcome the constraints of labeled data, Han et al. [7] introduce the task of few-shot relation extraction (FSRE), which enables models to handle RE task with a handful number of support instances.

Supported by National Key R&D Program of China (2021ZD0113902).

One of the prevalent approach for few-shot relation extraction (FSRE) is the Prototypical Network [15], which is rooted in the meta-learning framework [8] and aims to generalize to new classes not seen in the training set. Specifically, Prototypical Network learns a metric space in which classification can be performed by computing distances to prototype representation of each class. In the context of FSRE tasks, the prototype representation for each relation class is calculated by averaging the relation embeddings of support instances. Then the network is optimized by minimizing the distances between query samples and their corresponding relation prototypes.

To enhance the representation of relation prototypes, several studies have incorporated external knowledge, such as relation labels, relation descriptions and entity descriptions. TD-proto [17] enhances the prototypical network through a collaborative attention module, which allows to capture relevant information from the descriptions of entity and relation. CTEG [16] introduces an Entity-Guided Attention mechanism to alleviate the relation confusion by incorporating the syntactic relations and relative positions between each word and the specified entity pair. MTB [1] and CP [13] propose contrastive pre-training frameworks to obtain a relation representation with deeper understanding.

However, we argue that there are still two limitations for the calculation of relation prototypes in existing works. Firstly, the relation prototypes are primarily calculated based on the relation representations of support instances. The conventional approaches, inspired by MTB [1], concatenate the embeddings of the start tokens of the entity pair to obtain the relation representation of each support instance. Intuitively, the approach of relation representation fails to fully capture semantic nuances within complex scenes, where the identical entity pairs often convey diverse semantic relationship. Furthermore, when the support set has adequate labeled instances, the relation prototypes can be exclusively acquired from the support set. At this point, the direct introduction of additional relation knowledge potentially leads to instability of the relation prototypes.

Inspired by the success of prompt-tuning [10], in this paper, we propose a novel relation representation for support examples by utilizing a prompt template. Prompt learning fuses the original input with a prompt template to predict [MASK] and then maps the predicted label words to the respective class sets, which has induced better performances on few-shot tasks. Specifically, we augment the context of each input instance by inserting entity markers around entity pairs and appending a plain prompt template at the end of sentence. Subsequently, we concatenate the embeddings of the [CLS] and [MASK] token to obtain the relation representation of each instance. Finally, we deploy a gate mechanism [3] to dynamically select external relation representation originating from the relation name and description. The external relation representation is generated in accordance with the two relation views (i.e., [CLS] token embedding and the mean value of embeddings of all tokens) introduced by SimpleFSRE [11].

Figure 1 provides an intuitive illustration of the difference in ways to calculate relation representation of instance between existing works and our proposed approach. The conventional approach hinges upon shallow heuristics from entity mentions and cannot well understand the context. The observation motivates us to further improve the reliability of relation representation by refraining from mere memorization of entity names. We conduct experiments on the widely-used few-shot relation extraction datasets: FewRel 1.0 [7] and FewRel 2.0 [5]. The experimental results demonstrate the competitiveness of our approach compared to the state-of-the-art and significant improvements are observed in few-shot domain adaptation task. Our code is available at https://github.com/bebujoie/PNPT. Main results in this paper can be found in the CodaLab competition (bebu) at https://codalab.lisn.upsaclay.fr/competitions/7395#results and https://codalab.lisn.upsaclay.fr/competitions/7397#results.

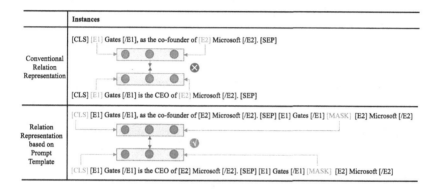

Fig. 1. The figure is an intuitive illustration of the differences in obtaining relation representations. The conventional approaches concatenate the embeddings of the start tokens of the entity pair to obtain the relation representation of input sentence. Our proposed method calculates the representation by concatenating the embeddings of the special tokens [CLS] and [MASK]. Compared to the former, our approach can avoid simple memorization for entity names.

The contributions of this paper are summarized as follows:

- We propose a novel relation representation for entity pairs based on a prompt template. Our approach is more expressive by integrating both the contextual and the entity mentions information.
- We adopt gate mechanism to selectively merge the external relation information. The approach can dynamically enhance the original relation prototypes according to the number of support instances.
- Experimental results on two FSRE benchmarks demonstrate the effectiveness of our approach. Specially, significant improvements are observed in few-shot domain adaptation task.

2   Approach

In this section, we present the details of proposed model for few-shot relation extraction. The overall architecture is illustrated in Fig. 2. In *N-way-K-shot* relation extraction task, a support set $\mathcal{S}$ and a query set $\mathcal{Q}$ are sampled from an auxiliary dataset. The support set $\mathcal{S}$ contains N classes, each with K labeled samples. The query set $\mathcal{Q}$ includes the same N classes as $\mathcal{S}$. Firstly, we calculate the initial prototype for each relation class according to relevant K support instances. Meanwhile, we can also acquire an additional relation representation by encoding relation name and description. To obtain the final relation prototype, we utilize a gate mechanism to dynamically fuse the original prototype and additional relation representation. When predicting the classification of instances within the query set, we compute the distances to prototype representation of each class.

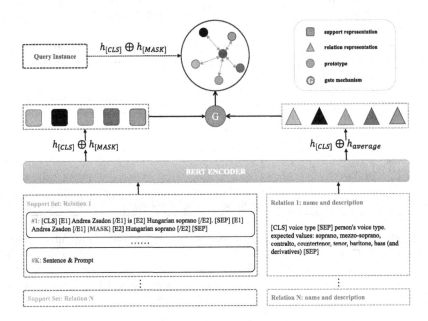

Fig. 2. The overall framework of our approach. A shared pre-trained BERT is deployed to generate original relation prototypes and external representation from support instances and relation information respectively. The rectangles represent the original prototypes, the triangle represent the external representation, the circles represent the final prototypes, and different colors represent different classes. $\oplus$ denotes the concatenation operation.

2.1 Prompt Function

We define a prompting function $f_{\mathrm{prompt}}(x)$ that transforms the original input into an augmented context. Specifically, we introduce four special tokens [E1] and [/E1], [E2] and [/E2] as entity markers, which serve to mark the begin and end of head and tail entities in the relation statement, respectively. Subsequently, we append a simple prompt template to the end of sentence. For a given sentence

$X = \{x_0 \ldots x_n\}$ with an entity pair $e_1 = \{x_i \ldots x_j\}$ and $e_2 = \{x_l \ldots x_m\}$, we modify X as follows:

$$\tilde{X} = \{[CLS]\, x_0 \cdots [E1]\, x_i \cdots x_j\, [/E1] \cdots [E2]\, x_l \cdots x_m\, [/E2] \cdots x_n\, [SEP]$$
$$[E1]\, x_i \cdots x_j\, [/E1]\, [MASK]\, [E2]\, x_l \cdots x_m\, [/E2]\, [SEP]\}, \tag{1}$$

where entity markers are inserted around entity e_1 and e_2, with a simple prompt appended to the end.

2.2 Sentence Encoder

To acquire contextualized embeddings of enhanced support instances and relation information, we employ the pre-trained BERT [4] as the shared encoder. Unlike the conventional relation representation, our proposed approach concatenates the embeddings of the special tokens [CLS] and [MASK] as the relation representation of input instance $\tilde{X}$, namely:

$$e_{\tilde{X}} = h_{[CLS]} \oplus h_{[MASK]}, \tag{2}$$

where $h_{[CLS]}, h_{[MASK]} \in \mathbb{R}^d$, denote the embeddings of token [CLS] and [MASK] contained in augmented input; $\oplus$ denotes concatenation operation and d is the size of the encoder hidden state.

For each relation, we directly concatenate its name and description and feed the sequence into the BERT encoder. Then the external relation representation is obtained through the two views method [11], where the embedding of the token [CLS] and the average of the embeddings of all tokens as two different views of relation representation:

$$e_{r_i} = \mathcal{R}_i^{view1} \oplus \mathcal{R}_i^{view2}, \tag{3}$$

where $\mathcal{R}_i^{view1} = h_{[CLS]}, \mathcal{R}_i^{view2} = \text{average}(h_j) \in \mathbb{R}^d\ (i = 1, 2, \cdots, N)$ represent two views respectively, $\oplus$ denotes concatenation operation and d is the size of the encoder hidden state.

2.3 Prototype Calculation

In the support set $\mathcal{S}$ of the N-way-K-shot task, we have N relation classes, each with K instances. We calculate the initial relation prototype for each class by averaging the relation representations of K instances:

$$\mathcal{P}_i = \frac{1}{K} \sum_{k=1}^{K} e_{\tilde{S}_k} \in \mathbb{R}^{2d}, \tag{4}$$

where $\mathcal{P}_i(i = 1, \cdots, N)$ denotes the initial relation prototype for class i.

The introduction of external relation information can make the prototypes more stable and distinctive. Therefore, we apply a gate mechanism [3] to calculate the final relation prototypes, as follows:

$$g = sigmoid(W_{pg}\mathcal{P}_i + b_{pg} + W_{rg}e_{r_i} + b_{rg}) \tag{5}$$

$$\hat{\mathcal{P}}_i = g \odot \mathcal{P}_i + (1-g) \odot e_{r_i}, \tag{6}$$

where $W_{pg}, W_{rg} \in \mathbb{R}^{2d \times 2d}, b_{pg}, b_{rg} \in \mathbb{R}^{2d \times 1}$ is trainable parameters, $\odot$ denotes element-wise multiplication, and $\hat{\mathcal{P}}_i$ represent the final prototype for relation i.

2.4   Relation Classification

Our approach utilizes the Euclidean distance as the metric for assessing similarity, while the cross-entropy (CE) is adopted as the loss function. During the inference stage, the model calculates the Euclidean distance between the representation of query instance and each relation prototype, and selects the closest as the predicted label:

$$e_q = h_{[CLS]} \oplus h_{[MASK]} \tag{7}$$

$$d(e_q, \hat{\mathcal{P}}_i) = \|e_q - \hat{\mathcal{P}}_i\|^2 \tag{8}$$

$$p(y_i|q) = \frac{\exp(-d(e_q, \hat{\mathcal{P}}_i))}{\sum_k \exp(-d(e_q, \hat{\mathcal{P}}_k))} \tag{9}$$

$$\mathcal{L}_{CE} = -\frac{1}{|\mathcal{Q}|} \sum_{q \in \mathcal{Q}} \sum_{i=1}^{N} p(y_i|q) \log p(y_i|q), \tag{10}$$

where e_q denotes the relation representation of query instance, $|\mathcal{Q}|$ denotes the numbers of query set, and N denotes the number of classes.

3   Experiments

3.1   Experiment Settings

To evaluate the effectiveness of our approach, we conduct three comparative trials as follows:

- **The Prototypical Network relies solely on support instances.** In this trial, we construct a prototypical network based on the support sets without external relation information. The BERT-base-uncased model [4] is harnessed as the encoder and the original prototype $\mathcal{P}_i$ directly serves as prototype of each relation class.
- **The enhanced Prototypical Network with relation information.** We enhance the original prototype of each relation class by incorporating the external information from relation name and description. Similar to the first experiment, we employ the BERT-base-uncased model [4] as the shared encoder to generate the final relation prototype $\hat{\mathcal{P}}_i$.

- **Contrastive pre-trained relation extraction model as encoder.** In this situation, we utilize the CP [13] as the encoder to obtain contextualized embedding of the input. CP is further pre-trained model based on BERT for relation extraction tasks and can achieve a deeper understanding on textual context through an entity-masked contrastive pre-training.

Datasets. Our proposed approach is evaluated on FewRel 1.0 [7] and FewRel 2.0 [5]. The FewRel 1.0 focuses on few-shot relation extraction (RE) and consists of 100 relations, with each relation having 700 labeled instances. To ensure a fair comparison, we adopt the same data splits as the official benchmarks, where 64 relation classes for training, 16 classes for validation and 20 classes for testing. FewRel 2.0 introduces a few-shot domain adaptation task by constructing new validation and test sets cover the biomedical domain. Meanwhile, the training set remains consistent with that of FewRel 1.0. The detailed statistics of the datasets are presented in Table 1.

Table 1. Statistics of FewRel 1.0 and FewRel 2.0 benchmark.

Dataset	Split	Relation	Instance
Common	Training	64	44800
FewRel 1.0	Validation	16	11200
	Test	20	14000
FewRel 2.0	Validation	10	1000
	Test	15	1500

Training. In all experimental settings, we ensure consistent training parameters except for learning rate. Specifically, we set the number of training iterations as 30,000, the number of validation iterations as 1,000 and batch size as 4. We employ the AdamW optimizer to optimize our model and set the initial learning rate as 1e−5 and 5e−6 for BERT and CP respectively. We also employ linear learning rate decay and early stopping strategies. The former facilitates smoother convergence by gradually reducing the learning rate over time. The latter prevents overfitting by monitoring the averaged accuracy on the validation set and terminating training if the performance does not improve significantly.

Evaluation. Consistent with the official benchmark evaluation method, our study focuses on *N-way-K-shot* relation extraction (RE) tasks. Here, N represents the number of relation classes, and K represents the number of instances from each class in a training episode. The performance of the model is evaluated as the averaged accuracy on the query sets. We select the model that demonstrates the highest performance on the validation set to inference on the test set. Given that the labels of the test set in FewRel are not publicly available, we report the final test results by submitting the prediction of our model to the CodaLab platform.

Baselines. We compare our model with several strong baseline approaches, including Proto-BERT [15], BERT PAIR [5], REGRAB [14], TD-proto [17], CTEG [16], Concept-FERE [18], MTB [1], CP [13], HCRP [6] and SimpleFSRE [11]. The majority of the aforementioned methods are evaluated only on FewRel 1.0. In light of this, we select several representative models for the purpose of comparison on FewRel 2.0, rather than encompassing all available options.

3.2   Results

Results on FewRel 1.0 Dataset. Table 2 presents the experimental results on FewRel 1.0 validation set and test set. The results are categorized into two sections based on the encoder utilized within our framework. In the top section, the pre-trained BERT is deployed as the encoder. It can be observed that our proposed approach achieves state-of-the-art in *5-way-5-shot* and *10-way-5-shot* settings, surpassing the baselines even without additional relation information. Furthermore, by integrating relation name and description through a gate mechanism, our approach achieves competitive results across all settings, outperforming the state-of-the-art except for the *5-way-1-shot* scenario. For fair comparison, we also employ the CP model as the backend encoder. Although the relation representation adopted in CP differs from our approach, the results demonstrate a further improvement in classification accuracy and achieve the state-of-the-art apart from the *10-way-1-shot* setting. These observations serve as evidence for the effectiveness of our proposed approach in few-shot relation extraction task.

Results on FewRel 2.0 Dataset. Table 3 presents the experimental results on FewRel 2.0 test set. FewRel 2.0 introduces a challenging task, evaluating the capacity of model to transfer knowledge across varying domains. Simultaneously, it provides fewer external relation information in comparison to FewRel 1.0, with only the relation name available for reference. The results demonstrate our method outperforms the strong baseline models and achieves the highest performance across all settings on FewRel 2.0. Particularly, we improve the averaged accuracy 3.61 points compared to the second-best approach. The remarkable improvement signifies the superior domain adaptation capability of our proposed approach. The performance gain mainly comes from two aspects: (1) The relation representation based on prompt template captures rich and subtle context features contained in support instances. (2) The gate mechanism dynamically integrates the relation name to further obtain discriminative prototypes.

Comparison with Proto-BERT. In this part, we present a direct comparison between our proposed approach and Proto-BERT under the same conditions,

Table 2. Accuracy (%) of few-shot classification on the FewRel 1.0 validation/test set, where *N-w-K-s* stands for the abbreviation of *N-way-K-shot*. The table is divided in two parts according to encoder, from top to bottom including approaches with pre-trained BERT and CP. Note that * is reported by SimpleFSRE [11] and others are obtained from papers or CodaLab.

Model	5-w-1-s	5-w-5-s	10-w-1-s	10-w-5-s	Avg.
Proto-BERT*	84.77/89.33	89.54/94.13	76.85/83.41	83.42/90.25	83.65/89.28
BERT-PAIR	85.66/88.32	89.48/93.22	76.84/80.63	81.76/87.02	83.44/87.28
REGRAB	87.95/90.30	92.54/94.25	80.26/84.09	86.72/89.93	86.87/89.64
TD-proto	−−/84.76	−−/92.38	−−/74.32	−−/85.92	−−/84.35
CTEG	84.72/88.11	92.52/95.25	76.01/81.29	84.89/91.33	84.54/89.00
ConceptFERE	−−/89.21	−−/90.34	−−/75.72	−−/81.82	−−/84.27
HCRP	90.90/93.76	93.22/95.66	84.11/89.95	87.79/92.10	89.01/92.87
SimpleFSRE	91.29/**94.42**	94.05/96.37	86.09/90.73	89.68/93.47	90.28/93.75
Ours(w/o rel)	87.24/89.69	94.17/96.58	78.92/84.41	88.79/93.52	87.28/91.05
Ours(w/rel)	92.49/93.77	94.30/**96.79**	84.73/**90.80**	88.76/**94.03**	90.70/**93.85**
MTB	−−/91.10	−−/95.40	−−/84.30	−−/91.80	−−/90.65
CP	−−/95.10	−−/97.10	−−/91.20	−−/94.70	−−/94.53
HCRP(CP)	94.10/96.42	96.05/97.96	89.13/93.97	93.10/96.46	93.09/96.20
SimpleFSRE(CP)	96.21/96.63	97.07/97.93	93.38/**94.94**	95.11/96.39	95.44/96.47
Ours(CP)	96.31/**96.99**	97.82/**98.28**	92.65/94.41	95.58/**96.68**	95.59/**96.59**

Table 3. Accuracy (%) of few-shot classification on the FewRel 2.0 domain adaptation test set, where *N-w-K-s* stands for the abbreviation of *N-way-K-shot*. * are reported by the paper [6], ⋆ are quoted from the paper [13] and ♣ represents the results of our implementation.

Model	5-w-1-s	5-w-5-s	10-w-1-s	10-w-5-s	Avg.
Proto-BERT*	74.60	82.70	63.50	76.50	74.33
BERT-PAIR*	67.41	78.57	54.89	66.85	66.93
MTB*	74.70	87.90	62.50	81.10	76.55
CP*	79.70	84.90	68.10	79.80	78.18
HCRP*	76.34	83.03	63.77	72.94	74.02
SimpleFSRE♣	74.47	88.44	61.87	76.01	75.20
Ours	**81.81**	**90.67**	**70.30**	**84.40**	**81.79**

where both methods rely solely on support instances to construct Prototypical Network without utilizing additional relation information. The primary distinction between the two methods is the relation representation of support instances. Proto-BERT acquires the relation representation by concatenating entity marker embeddings corresponding to head and tail entities. Specifically, for a sentence containing entities e_1 and e_2, the relation representation between the entity pair is obtained by inserting special entity markers, resulting in $x = [h_{[E1]}; h_{[E2]}]$. Our approach generates the relation representation by utilizing a prompt template, namely $x = [h_{[CLS]}; h_{[MASK]}]$. The experimental results shown in Table 2 reveal that our approach outperforms Proto-BERT in all settings, particularly when the K is set to 5. The comparison with Proto-BERT validates the effectiveness of our approach and emphasizes the significance of relation representation based on prompt template.

3.3 Ablation Study

In this section, we investigate the impact of the individual representation by performing an ablation study on the FewRel 1.0. Due to the unavailability of test set labels in FewRel 1.0, our ablation study is conducted on the validation set with the pre-trained BERT. The results of our study are presented in Fig. 3.

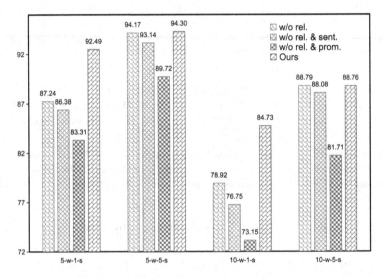

Fig. 3. Ablation study on validation set of FewRel 1.0, where w/o, rel., sent. and prom. are the abbreviations of without, relation representation, sentence representation and prompt mask representation.

Effect of External Relation Information. The first ablation study is "w/o rel.", which investigates the impact of excluding additional relation information in the calculation of prototypes. In this study, the prototypes are derived solely from the sentence representation and mask representation of support instances, without incorporating any relation-specific information. The results indicate a noticeable decrease in averaged accuracy, with values dropping from 92.49% and 84.73% to 87.24% and 78.92% in the *5-way-1-shot* and *10-way-1-shot* scenarios, respectively. These findings underscore the criticality of introducing relation information in classification tasks that involve single instance.

Effect of Sentence Global Representation. In this ablation study, we focus on calculating the relation prototype exclusively based on the mask representation within the prompt template whereas neglecting both the external relation information and the global representation of instances. Notably, we observe that the performance shows no significant decline across all settings when compared to the "w/o rel." condition. Therefore, while the sentence global representation can provide supplementary insights, it is not a pivotal component in constructing the prototypical network for few-shot relation classification task.

Effect of Mask Representation Within Prompt Template. Furthermore, we conduct an additional ablation study to examine the influence of the prompt mask representation. The results reveal a notable decline in all situations when the prompt mask representation was removed. This observation emphasizes the vital function of the mask representation in constructing the prototypical network. It demonstrates that the mask representation effectively captures the essential characteristics of the relation between entity pairs, particularly when utilized within an appropriate prompt template.

Overall, our comprehensive analysis indicates that all modules employed in our approach significantly contribute to the overall performance. Notably, the relation information module exhibits higher sensitivity to performance, particularly in the *1-shot* scenario. Moreover, we find that the prompt mask representation module plays a crucial and indispensable role, particularly when the value of K is set to 5. The prompt mask representation can effectively capture the essential characteristics of relations, enabling the model to better understand and inference.

4   Conclusion

In this paper, we propose a novel approach for few-shot relation extraction that addresses the limitations of existing prototypical network-based methods. Our approach focuses on enhancing the relation representation of support instances by utilizing a prompt template, resulting in a more informative and expressive relation prototype. To effectively integrate additional relation knowledge without causing conflicts with the support instance representation, we have introduced a gate mechanism. The mechanism dynamically merges external knowledge with the initial relation prototype, allowing for a more comprehensive and accurate prototype representation of each relation class. Experimental results on two benchmark demonstrate the effectiveness of our approach. The proposed method achieves competitive performance compared to state-of-the-art baselines. We will consider to design a mechanism for generating dynamical prompt template rather than a simple static template in our future work.

Acknowledgments. This work is supported by National Key R&D Program of China (2021ZD0113902).

References

1. Baldini Soares, L., FitzGerald, N., Ling, J., Kwiatkowski, T.: Matching the blanks: distributional similarity for relation learning. In: Proceedings of the 57th Annual Meeting of the Association for Computational Linguistics, pp. 2895–2905. Association for Computational Linguistics, Florence, Italy, July 2019. https://doi.org/10.18653/v1/P19-1279
2. Bordes, A., Chopra, S., Weston, J.: Question answering with subgraph embeddings. In: Proceedings of the 2014 Conference on Empirical Methods in Natural Language Processing (EMNLP), pp. 615–620. Association for Computational Linguistics, Doha, Qatar, October 2014. https://doi.org/10.3115/v1/D14-1067
3. Cho, K., van Merriënboer, B., Bahdanau, D., Bengio, Y.: On the properties of neural machine translation: Encoder-decoder approaches. In: Proceedings of SSST-8, Eighth Workshop on Syntax, Semantics and Structure in Statistical Translation, pp. 103–111. Association for Computational Linguistics, Doha, Qatar, October 2014. https://doi.org/10.3115/v1/W14-4012
4. Devlin, J., Chang, M.W., Lee, K., Toutanova, K.: BERT: pre-training of deep bidirectional transformers for language understanding. In: Proceedings of the 2019 Conference of the North American Chapter of the Association for Computational Linguistics: Human Language Technologies, Volume 1 (Long and Short Papers), pp. 4171–4186. Association for Computational Linguistics, Minneapolis, Minnesota, Jun 2019. https://doi.org/10.18653/v1/N19-1423
5. Gao, T., et al.: FewRel 2.0: Towards more challenging few-shot relation classification. In: Proceedings of the 2019 Conference on Empirical Methods in Natural Language Processing and the 9th International Joint Conference on Natural Language Processing (EMNLP-IJCNLP), pp. 6250–6255. Association for Computational Linguistics, Hong Kong, China, November 2019. https://doi.org/10.18653/v1/D19-1649
6. Han, J., Cheng, B., Lu, W.: Exploring task difficulty for few-shot relation extraction. In: Proceedings of the 2021 Conference on Empirical Methods in Natural Language Processing, pp. 2605–2616. Association for Computational Linguistics, Online and Punta Cana, Dominican Republic, November 2021. https://doi.org/10.18653/v1/2021.emnlp-main.204
7. Han, X., et al.: FewRel: a large-scale supervised few-shot relation classification dataset with state-of-the-art evaluation. In: Proceedings of the 2018 Conference on Empirical Methods in Natural Language Processing, pp. 4803–4809. Association for Computational Linguistics, Brussels, Belgium, October-November 2018. https://doi.org/10.18653/v1/D18-1514
8. Hospedales, T., Antoniou, A., Micaelli, P., Storkey, A.: Meta-learning in neural networks: a survey. IEEE Trans. Pattern Anal. Mach. Intell. **44**(9), 5149–5169 (2021). https://doi.org/10.1109/TPAMI.2021.3079209
9. Li, Z., Liu, H., Zhang, Z., Liu, T., Xiong, N.N.: Learning knowledge graph embedding with heterogeneous relation attention networks. IEEE Trans. Neural Netw. Learn. Syst. **33**(8), 3961–3973 (2022). https://doi.org/10.1109/TNNLS.2021.3055147
10. Liu, P., Yuan, W., Fu, J., Jiang, Z., Hayashi, H., Neubig, G.: Pre-train, prompt, and predict: a systematic survey of prompting methods in natural language processing. ACM Comput. Surv. **55**(9) (2023). https://doi.org/10.1145/3560815

11. Liu, Y., Hu, J., Wan, X., Chang, T.H.: A simple yet effective relation information guided approach for few-shot relation extraction. In: Findings of the Association for Computational Linguistics: ACL 2022, pp. 757–763. Association for Computational Linguistics, Dublin, Ireland, May 2022. https://doi.org/10.18653/v1/2022.findings-acl.62

12. Pawar, S., Palshikar, G.K., Bhattacharyya, P.: Relation extraction: a survey. arXiv preprint arXiv:1712.05191 (2017)

13. Peng, H., et al.: Learning from context or names? An empirical study on neural relation extraction. In: Proceedings of the 2020 Conference on Empirical Methods in Natural Language Processing (EMNLP), pp. 3661–3672. Association for Computational Linguistics, Online, November 2020. https://doi.org/10.18653/v1/2020.emnlp-main.298

14. Qu, M., Gao, T., Xhonneux, L.P.A.C., Tang, J.: Few-shot relation extraction via Bayesian meta-learning on relation graphs. In: Proceedings of the 37th International Conference on Machine Learning. ICML'20, JMLR.org (2020)

15. Snell, J., Swersky, K., Zemel, R.: Prototypical networks for few-shot learning. In: Proceedings of the 31st International Conference on Neural Information Processing Systems. NIPS'17, pp. 4080–4090. Curran Associates Inc., Red Hook, NY, USA (2017)

16. Wang, Y., et al.: Learning to decouple relations: few-shot relation classification with entity-guided attention and confusion-aware training. In: Proceedings of the 28th International Conference on Computational Linguistics, pp. 5799–5809. International Committee on Computational Linguistics, Barcelona, Spain (Online), December 2020. https://doi.org/10.18653/v1/2020.coling-main.510

17. Yang, K., Zheng, N., Dai, X., He, L., Huang, S., Chen, J.: Enhance prototypical network with text descriptions for few-shot relation classification. In: Proceedings of the 29th ACM International Conference on Information and Knowledge Management. CIKM '20, pp. 2273–2276. Association for Computing Machinery, New York, NY, USA (2020). https://doi.org/10.1145/3340531.3412153

18. Yang, S., Zhang, Y., Niu, G., Zhao, Q., Pu, S.: Entity concept-enhanced few-shot relation extraction. In: Proceedings of the 59th Annual Meeting of the Association for Computational Linguistics and the 11th International Joint Conference on Natural Language Processing (Volume 2: Short Papers), pp. 987–991. Association for Computational Linguistics, Online, August 2021. https://doi.org/10.18653/v1/2021.acl-short.124

CDBMA: Community Detection in Heterogeneous Networks Based on Multi-attention Mechanism

Yuanxin Li[1,2] , Zhixiang Wu[1(✉)] , Zhenyu Wang[1] , and Ping Li[1]

[1] School of Software Engineering, South China University of Technology,
Guangzhou, China
liyuanxin@youcash.com,
{202021045997,wangzy,202211088875}@mail.scut.edu.cn
[2] PSBC Consumer Finance, Guangzhou, China

Abstract. Community detection in complex networks is a fundamental task in network analysis. With the continuous evolution of social networks, network structures are becoming more complex and often contain rich heterogeneous information. Traditional community detection methods can only utilize shallow topological features and fail to leverage the rich heterogeneous information in these networks, making community detection in heterogeneous networks a new challenge. In this paper, we propose a community detection model for heterogeneous networks based on multi-attention mechanisms. Our model consists of a structural information encoder and a semantic information encoder. The structural information encoder proposes a subgraph sampler to extract subgraphs around target nodes, uses type attention to aggregate the influence of different types of nodes, and learns the heterogeneous structural information of the network. The semantic information encoder uses node attention to learn the importance of high-order neighbor nodes based on meta-paths, uses semantic attention to learn the weights of different meta-paths, and fuses the content semantic information on different meta-paths to learn the content semantic information of the heterogeneous network. Joint optimization of structural and semantic encoders is achieved through self-supervised learning, addressing the dependence on community labels. We evaluate our model on four real-world datasets, and the results show that our algorithm outperforms several community detection state-of-the-art methods, especially approximate 10% improvement in the NMI metric on the ACM and Freebase datasets, and approximate 20% improvement in the ARI metric on the ACM dataset.

Keywords: Community detection · Graph neural network · Heterogeneous network

1 Introduction

With the emergence of 5G networks and the continuous development of new media technologies, online social networks have played an important role in

F. Wu et al. (Eds.): SMP 2023, CCIS 1945, pp. 174–187, 2024.
https://doi.org/10.1007/978-981-99-7596-9_13

sharing and disseminating information. Social software and platforms such as WeChat, QQ, Facebook, and Twitter facilitate daily communication and interaction among people. More and more users share various information through social platforms, and social networking is gradually transitioning from offline to online. Online social networks have changed the way people communicate and their habits, creating a virtual network world among real-world people. Users in the network are real individuals, and information flows and spreads in the virtual network world, which can have a significant impact on people's lives in the real world, as well as market behavior. Therefore, studying and analyzing social networks is crucial, with significant research significance and value, and can be applied to various fields such as information dissemination, personalized recommendation, and network marketing.

Community structure, as an important structural feature, is used for the decomposition and partitioning of social networks. Users within the same community share similar interests and hobbies, while users from different communities have different interests and hobbies. Community detection in social networks can effectively provide personalized recommendations [23,24] for users and monitor public opinion on information dissemination, with significant research value. With the continuous evolution of social networks, network structures are becoming more complex and often contain rich heterogeneous information, showing high-dimensional and sparse data characteristics. Traditional community detection methods can only utilize shallow topological features to discover communities [1,25].

In this paper, we conducted research on community detection in heterogeneous networks based on graph neural network technology and proposed a multi-attention mechanism-based model for community detection in heterogeneous networks.

In summary, this paper made the following contributions:

- We proposed a multi-attention mechanism-based model for community detection in heterogeneous networks, which includes a structural information encoder and a semantic information encoder to simultaneously learn the heterogeneous structural information and content semantic information of the network.
- We used multiple attention mechanisms, including type attention to aggregate the influence of different types of nodes, node attention to learn the importance of high-order neighbor nodes based on meta-paths, and semantic attention to learn the weights of different meta-paths.
- We jointly optimized the structural information encoder and the semantic information encoder through self-supervised learning to enable mutual benefits between modules and address the dependence on community labels.
- Our algorithm showed state-of-the-art performance in community detection compared to other baseline techniques.

2   Related Work

Based on deep learning methods, the ability to consider network heterogeneity or semantic information can be improved for community detection. Chen et al. [1] proposed a linear neural network model LGNN, which uses a semi-supervised learning approach based on graph neural networks to model the community detection problem. Hu et al. [2] proposed a heterogeneous network model based on Transformer, which uses parameters related to node and edge types to characterize heterogeneous attention on each edge.

Meta-paths provide a systematic way to define context in heterogeneous networks [3]. Heterogeneous graph embedding based on meta-paths can capture high-order relationships between any two nodes in the network. Wang et al. [4] proposed the Heterogeneous Graph Attention Model (HAN), which aggregates neighbor node information from different meta-paths hierarchically to learn representations for heterogeneous network nodes, including node attention and semantic attention. This method only aggregates semantic information from meta-paths and requires prior knowledge of the true community structure. In contrast, Zhang et al. [5] used LSTM to aggregate node-level information and attention mechanisms to aggregate semantic-level information, resulting in node representations for clustering.

Existing meta-path-based heterogeneous representation learning models often ignore node content information features and only consider information between rich nodes within a single meta-path. To address this, Fu et al. [6] proposed a novel meta-path aggregated heterogeneous network representation model MAGNN, which uses two stages of meta-path aggregation to model the heterogeneous network, capturing rich structural and semantic information, then applies the learned embeddings to community detection. Fang et al. [7] extended the classical minimum degree measure using meta-paths to measure community cohesion, and proposed an effective community search algorithm using these cohesion measures, but it cannot be directly used for community detection.

There are also works that avoid using pre-defined meta-paths. For example, Luo et al. [8] proposed a contextual path graph neural network model CP-GNN for detecting communities from heterogeneous graphs, which captures high-order relationships between nodes effectively. Wang et al. [9] proposed a new semantic and relation-aware heterogeneous graph neural network model, which uses semantic-aware and relation-aware attention to model the heterogeneous network, considering the heterogeneity on nodes and relations, but requires prior knowledge of the true community structure for supervised training. Wu et al. [10] proposed a graph neural network-based heterogeneous community detection algorithm HCDBG for discovering complex question-answering communities, which quantifies various entities and complex relationships in the question-answering system using a heterogeneous information network, and then uses heterogeneous graph neural networks for entity embedding and an improved K-means algorithm to discover hidden communities.

Some works have also been proposed for overlapping community detection in heterogeneous networks. For example, Yue et al. [11] proposed the HDPCOCD model, which combines the improved HAN model with density peak clustering for finding overlapping communities in heterogeneous networks.

3 Preliminary

In this section, we formally define some significant concepts related to Heterogeneous Network as follows:

Network Schema: A network schema $T_G = (\mathcal{A}, \mathcal{R})$ is a highly abstracted meta-template representation of the heterogeneous information network, which includes all node types $\mathcal{A}$ and relationship types $\mathcal{R}$ in the heterogeneous information network, typically represented as a directed graph.

Meta-path: A meta-path Φ is defined on a path pattern $S = (A, R)$ in the heterogeneous information network, represented as $A_1 \xrightarrow{R_1} A_2 \xrightarrow{R_2} \ldots \xrightarrow{R_l} A_{l+1}$, where $R = R_1 \circ R_2 \circ \cdots \circ R_l$ is a composite relationship defined on node object types $A_1, A_2, \cdots, A_{l+1}$, and $\circ$ represents a relationship composition operator.

Meta-path-Based Neighbor: In the heterogeneous information network, for a given node i, the meta-path-based neighbor $\mathcal{N}_i^\Phi$ is the set of nodes that are indirectly connected to node i through the meta-path Φ. Typically, this set of nodes also includes node i itself, i.e., each node is its own meta-path-based neighbor.

Problem Definition: Given a heterogeneous graph G, our goal is to learn a good target node embedding $\mathbf{Z} \in \mathbb{R}^{N \times d}$ where N denotes the number of target nodes and d is the embedding dimension, such that they can be used to group the nodes into a set of communities $C = \{1, \cdots, C\}$ with strong inner-connection.

4 Proposed Method

The overall framework of the model is shown in Fig. 1. The model consists of a Structure Information Encoder module (comprising Heterogeneous Subgraph Sampler, Same Type Node Aggregation Layer, and Different Type Node Aggregation Layer) and a Semantic Information Encoder module (utilizing Node Attention and Semantic Attention Hierarchical Aggregation). Through contrastive training, it combines heterogeneous structure and semantic information to learn representations of target nodes in a multi-modal heterogeneous network. Finally, the learned embeddings of target nodes are input to a K-means clustering layer to obtain community partitions of the target nodes.

4.1 Structure Information Encoder

Heterogeneous Subgraph Sampler. For a heterogeneous network $\mathcal{G} = (\mathcal{V}, \mathcal{E})$, the network schema $T_G = (\mathcal{A}, \mathcal{R})$ serves as the minimal subgraph of the heterogeneous network, containing all node types and relation types in $\mathcal{G}$. However, in multimodal heterogeneous networks, the number of neighboring nodes

178   Y. Li et al.

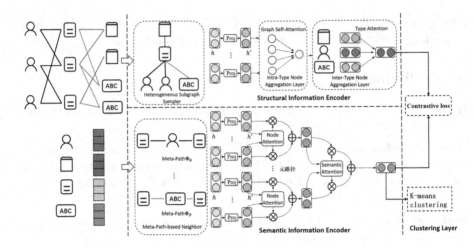

Fig. 1. Overall framework of CDBMA.

connected to a target node varies across different node types, potentially leading to an imbalanced distribution. To address this issue and prevent the dominance of a large number of nodes overshadowing the influence of other neighboring nodes, a heterogeneous subgraph sampler is introduced. This sampler selects a subset of heterogeneous neighboring nodes connected to the target node, with the number of sampled neighbors determined by the sample size. The sampler then returns the subgraph containing the target node.

Same-Type Node Aggregation Layer. In heterogeneous subgraphs, there exist heterogeneous neighboring nodes with features that are typically not in the same dimensional space. Therefore, it is necessary to preprocess the features of heterogeneous neighboring nodes and project them onto the same dimensional space:

$$\mathbf{h}'_i = \sigma\left(\mathbf{w}_{\phi_i} \cdot \mathbf{h}_i + \mathbf{b}_{\phi_i}\right) \tag{1}$$

In Eq. 1, $\mathbf{h}'_i$ represents the projected feature of node i, $\mathbf{h}_i$ represents the original feature of node i, $\sigma(\cdot)$ represents the activation function, $\mathbf{w}_{\phi_i}$ represents the mapping matrix of the corresponding node type ϕ_i, and $\mathbf{b}_{\phi_i}$ represents the bias vector.

In the sampled heterogeneous subgraph, suppose the set of neighbors connected to the target node i of type ϕ_m is denoted as $N_i^{\phi_m}$. For the target node i, even for different neighboring nodes of the same type, their contributions to the embedding of the target node are different. Therefore, we use a graph self-attention mechanism to aggregate the set of neighbors $N_i^{\phi_m}$ of the same type:

$$\mathbf{h}'^{\phi_m}_i = \sigma\left(\sum_{j \in N_i^{\phi_m}} \alpha^{\phi_m}_{i,j} \cdot \mathbf{h}'_j\right) \tag{2}$$

In Eq. 2, σ represents the nonlinear activation function, $\mathbf{h}'_j$ is the projected feature of node j, and $\alpha_{i,j}^{\phi_m}$ represents the attention coefficient of node j of type ϕ_m to the target node i, which is calculated as follows:

$$\alpha_{i,j}^{\phi_m} = \frac{\exp\left(\text{LeakyReLU}\left(\mathbf{a}_{\phi_m}^\top \cdot [\mathbf{h}'_i \| \mathbf{h}'_j]\right)\right)}{\sum_{l \in N_i^{\phi_m}} \exp\left(\text{LeakyReLU}\left(\mathbf{a}_{\phi_m}^\top \cdot [\mathbf{h}'_i \| \mathbf{h}'_l]\right)\right)} \tag{3}$$

In Eq. 3, LeakyReLU represents the activation function, $\mathbf{a}_{\phi_m} \in \mathbb{R}^{2d \times 1}$ represents the graph attention vector for a given neighbor type ϕ_m, $\|$ represents the matrix concatenation operation, and $N_i^{\phi_m}$ represents the set of neighbors connected to the target node i of type ϕ_m in the heterogeneous subgraph.

Different-Type Node Aggregation Layer. After aggregating the features of different neighbors of the same type for the target node in the heterogeneous subgraph, we obtain type embeddings from different types of neighbors $\left\{\mathbf{h}_i^{\prime\phi_1}, \mathbf{h}_i^{\prime\phi_2}, \ldots, \mathbf{h}_i^{\prime\phi_m}\right\}$.

For a target node i, the contributions of different neighbors of the same type and neighbors of different types to its embedding vary. After aggregating the features of same-type neighbors, the embedding for the target node is learned based on each neighbor type using type attention. These embeddings are then fused to obtain the final embedding for the target node. To distinguish between different types of heterogeneous neighbors, the importance of each type for the target node is learned and calculated accordingly.

$$w_{\phi_m} = \frac{1}{|v|} \sum_{i \in V} \mathbf{a}^\top \cdot \tanh\left(\mathbf{w}\mathbf{h}_i^{\prime\phi_m} + \mathbf{b}\right) \tag{4}$$

In formula 4, v represents the set of target nodes, $\mathbf{w} \in \mathbb{R}^{d \times d}$ represents the weight coefficients, $\mathbf{b} \in \mathbb{R}^{d \times 1}$ represents the bias vector, and $\mathbf{a}$ represents the type attention vector. After obtaining the importance of the heterogeneous neighbors w_{ϕ_m}, we further use the softmax function to normalize and obtain the weight coefficients β_{ϕ_m} of each type of heterogeneous neighbor for the target node, as calculated in the following formula:

$$\beta_{\phi_m} = \frac{\exp\left(w_{\phi_m}\right)}{\sum_{i=1}^S \exp\left(w_{\phi_i}\right)} \tag{5}$$

Finally, based on the learned weight coefficients, we fuse the type embeddings from different types of neighbors to obtain the embedding of the target node in the heterogeneous subgraph, as shown in the following formula:

$$\mathbf{z}_i^{sh} = \sum_{m=1}^S \beta_{\phi_m} \cdot \mathbf{h}_i^{\prime\phi_m} \tag{6}$$

4.2   Semantic Information Encoder

Node Attention. In heterogeneous networks, the feature dimensions of different types of nodes are mostly different. Therefore, it is necessary to preprocess the feature dimensions of different types of nodes and project them into the same dimensional space. For node i of type ϕ_i in the heterogeneous network, a specific type of transformation matrix $\mathbf{M}_{\phi_i}$ is designed to project the features of different types of heterogeneous nodes into the same dimensional space. The formula for transforming node feature dimensions is as follows:

$$\mathbf{h}'_i = \mathbf{M}_{\phi_i} \cdot \mathbf{h}_i \tag{7}$$

In Eq. 7, $\mathbf{h}'_i$ represents the projected feature of node i, and $\mathbf{h}_i$ represents the feature of node i before projection.

For neighbor node pairs (i, j) connected by a meta-path Φ, node attention can learn the importance of neighbor node j to node i based on the meta-path and calculate the formula as follows:

$$e^{\Phi}_{ij} = att_{node}\left(\mathbf{h}'_i, \mathbf{h}'_j; \Phi\right) \tag{8}$$

In Eq. 8, att_{node} represents a deep neural network that implements node attention. Nodes with a given meta-path Φ share the parameters of this neural network because the connection patterns of node pairs under the same meta-path exhibit similarity. The attention weight between neighbor node pairs (i, j) based on meta-path Φ depends on the features of the starting and ending nodes of the meta-path. This attention weight is asymmetric, reflecting the varying influence weights between neighbor nodes based on the meta-path and maintaining the graph's inherent asymmetry.

Node attention calculates the attention weight e_{ij} only between the target node and its neighbor nodes $j \in \mathcal{N}_i^{\Phi}$ based on the meta-path, rather than considering attention weights between all nodes. After obtaining the importance of the target node's neighbor nodes based on the meta-path, the softmax function is used to normalize and obtain the weight coefficient α_{ij}^{Φ}, following the calculation formula provided. Similar to Eq. 3.

To make the node attention learning more stable, Vaswani [26] proposed a multi-head attention mechanism that repeats the node attention K times and concatenates the learned embeddings of K target nodes to obtain the representation of the target node under a specific meta-path Φ. The calculation is shown as follows:

$$\mathbf{z}_i^{\Phi} = \Big\|_{k=1}^{K} \sigma\left(\sum_{j \in \mathcal{N}_i^{\Phi}} \alpha_{ij}^{\Phi} \cdot \mathbf{h}'_j\right) \tag{9}$$

For a given set of meta-paths $\{\Phi_1, \Phi_2, \ldots, \Phi_P\}$, applying node attention to the node features and meta-paths in the heterogeneous graph can obtain P sets of target node representations with specific semantic structures, denoted as $\left\{\mathbf{z}_i^{\Phi_1}, \mathbf{z}_i^{\Phi_2}, \ldots, \mathbf{z}_i^{\Phi_P}\right\}$.

Semantic Attention. The semantic information encoder takes the output of node attention $\left\{ \mathbf{z}_i^{\Phi_1}, \mathbf{z}_i^{\Phi_2}, \ldots, \mathbf{z}_i^{\Phi_P} \right\}$ as input to the next layer, and semantic attention can learn the weight $(\beta_{\Phi_1}, \beta_{\Phi_2}, \ldots, \beta_{\Phi_P})$ of different meta-paths as shown below:

$$(\beta_{\Phi_1}, \beta_{\Phi_2}, \ldots, \beta_{\Phi_P}) = \text{att}_{\text{semantic}} \left(\mathbf{z}_i^{\Phi_1}, \mathbf{z}_i^{\Phi_2}, \ldots, \mathbf{z}_i^{\Phi_P} \right) \tag{10}$$

In formula 10, $\text{att}_{\text{semantic}}$ represents a deep neural network that implements semantic attention. It can be seen that semantic attention distinguishes the differences in different types of semantic structure information in a multi-modal heterogeneous network and assigns suitable weights to them.

Semantic attention uses an MLP to map node representations to a shared feature space. The similarity between transformed node representations and the semantic attention vector determines meta-path importance. Importance w_{Φ_n} of each meta-path is obtained by fusing the importance of all nodes. The calculation is shown as follows:

$$w_{\Phi_n} = \frac{1}{|\mathcal{V}|} \sum_{i \in \mathcal{V}} \mathbf{q}^{\mathrm{T}} \cdot \tanh \left(\mathbf{w} \cdot \mathbf{z}_i^{\Phi_n} + \mathbf{b} \right) \tag{11}$$

In formula 11, $\mathbf{w}$ represents the weight coefficient, $\mathbf{b}$ represents the bias vector, $\mathbf{q}$ represents the semantic attention vector, and the parameters learned for meta-path weight in semantic attention are shared. The obtained importance w_Φ of meta-paths can be normalized using the Softmax function to obtain the normalized weight β_{Φ_i} of meta-paths, which is calculated as follows:

$$\beta_{\Phi_n} = \frac{\exp\left(w_{\Phi_n}\right)}{\sum_{i=1}^{P} \exp\left(w_{\Phi_i}\right)} \tag{12}$$

In formula 12, β_{Φ_i} can be understood as the contribution of the meta-path to the given task. It can be seen that the larger the weight β_{Φ_i} of the meta-path, the more important the meta-path Φ_i is for the given task. After obtaining the weight β_{Φ_i} of the meta-path, the node representations from different semantic structure information are fused by weighted summation according to the weight to obtain the representation of the target node under the meta-path, which is shown as follows:

$$\mathbf{z}_i^{mp} = \sum_{n=1}^{P} \beta_{\Phi_n} \cdot \mathbf{z}_i^{\Phi_n} \tag{13}$$

4.3   Joint Contrastive Optimization

Contrastive learning maximizes the similarity between positive samples and minimizes the similarity between negative samples to learn data representations. Contrastive optimization serves as a self-supervised method, allowing the structural and semantic encoders to supervise each other's training To compute the contrastive loss, after obtaining the embedding representations $\mathbf{z}_i^{sh}$ and $\mathbf{z}_i^{mp}$ of the target node from the two encoders, they are input into an MLP with one hidden layer to project them into the same space for computing the contrastive loss:

$$\mathbf{z}_i^{sh}\text{-}proj = \mathbf{W}^{(2)}\sigma\left(\mathbf{W}^{(1)}z_i^{sh} + \mathbf{b}^{(1)}\right) + \mathbf{b}^{(2)} \tag{14}$$

$$\mathbf{z}_i^{mp}\text{-}proj = \mathbf{W}^{(2)}\sigma\left(\mathbf{W}^{(1)}\mathbf{z}_i^{mp} + \mathbf{b}^{(1)}\right) + \mathbf{b}^{(2)} \tag{15}$$

In formula 15, σ represents the ELU activation function. Note that the embedding projection of the target node is shared among all MLPs, i.e., the parameters $\left\{\mathbf{W}^{(2)}, \mathbf{W}^{(1)}, \mathbf{b}^{(2)}, \mathbf{b}^{(1)}\right\}$ are shared. we adopt a novel positive sample selection strategy [15], where if two nodes are connected by multiple meta-paths, they are both positive samples for each other. This positive sample selection strategy can reflect the local structure of the target node.

For nodes i and j, we define a function $\mathbb{C}_i(\cdot)$ to compute the number of meta-paths connecting these two nodes as follows:

$$\mathbb{C}_i(j) = \sum_{n=1}^{M} \mathbb{K}\left(j \in N_i^{\mathcal{P}_n}\right) \tag{16}$$

where $\mathbb{K}(\cdot)$ is the indicator function. Then we construct a set $S_i = \{j \mid j \in V$ and $\mathbb{C}_i(j) \neq 0\}$ and sort it in the descending order based on the value of $\mathbb{C}_i(\cdot)$. Next we set a threshold T_{pos}, and if $|S_i| > T_{\text{pos}}$, we select first T_{pos} nodes from S_i as positive samples of i, denotes as $\mathbb{P}_i$, otherwise all nodes in S_i are retained. And we naturally treat all left nodes as negative samples of i, denotes as $\mathbb{N}_i$.

After selecting the positive sample set $\mathbb{P}_i$ and the negative sample set $\mathbb{N}_i$, the contrastive loss under the structural information encoder is as follows:

$$\mathcal{L}_i^{sh} = -\log \frac{\sum_{j \in \mathbb{P}_i} \exp\left(\text{sim}\left(\mathbf{z}_i^{sh}\text{-} proj, \mathbf{z}_j^{mp}\text{-}proj\right)/\tau\right)}{\sum_{k \in \{\mathbb{P}_i \cup \mathbb{N}_i\}} \exp\left(\text{sim}\left(\mathbf{z}_i^{sh}\text{-}proj, \mathbf{z}_k^{mp}\text{-}proj\right)/\tau\right)} \tag{17}$$

In Eq. 17, $\text{sim}(u, v)$ represents the cosine similarity between two vectors u and v, and τ is a temperature parameter. For a pair of nodes, the target embedding comes from the structural information encoder, while the positive and negative node embeddings come from the semantic information encoder, achieving self-supervised learning.

Comparative losses for both are similar, as shown in Eq. 18. However, the target embedding comes from the semantic information encoder, while the positive and negative node embeddings come from the structural information encoder.

$$\mathcal{L}_i^{mp} = -\log \frac{\sum_{j \in \mathbb{P}_i} \exp\left(\text{sim}\left(\mathbf{z}_i^{mp}\text{-} proj, \mathbf{z}_j^{sh}\text{-}proj\right)/\tau\right)}{\sum_{k \in \{\mathbb{P}_i \cup \mathbb{N}_i\}} \exp\left(\text{sim}\left(\mathbf{z}_i^{mp}\text{-}proj, \mathbf{z}_k^{sh}\text{-}proj\right)/\tau\right)} \tag{18}$$

By jointly optimizing the contrastive losses under the structural information encoder and the semantic information encoder, the model can learn both the heterogeneous structural information and content semantic information of the multi-modal heterogeneous network. The overall contrastive loss function of the model is as follows:

$$\mathcal{L} = \frac{1}{|V|} \sum_{i \in V} \left[\lambda \cdot \mathcal{L}_i^{sh} + (1 - \lambda) \cdot \mathcal{L}_i^{mp} \right] \qquad (19)$$

In Eq. 19, $\lambda > 0$ is a hyperparameter used to balance the contrastive losses under the structural information encoder and the semantic information encoder.

The model is trained using contrastive loss, and the learning of the target embedding is complemented by mutual supervision during the training process. Since the nodes of the target type explicitly participate in the generation of the target node embedding representation $\mathbf{z}_i^{mp}$, $\mathbf{z}_i^{mp}$ can be used for downstream tasks. We apply the K-means clustering algorithm to $\mathbf{z}_i^{mp}$ for community detection.

4.4   Experiment

Dataset. We used four real heterogeneous network datasets, namely ACM [12], Aminer [13], DBLP [6], and Freebase [14]. Each dataset contains its own heterogeneous structure and different meta-paths. The statistics of these four heterogeneous network datasets are shown in [15].

Baseline Methods. We compared nine algorithms with our method in experiments. These algorithms can be divided into four categories based on the types of network objects and training modes targeted by the algorithms: unsupervised homogeneous methods{GraphSAGE [16], GAE [17], DGI [18]}, unsupervised heterogeneous methods{Metapath2vec [19], HERec [20], HetGNN [5], DMGI [21]}, semi-supervised heterogeneous methods{HAN [4], HGT [22]}, and self-supervised heterogeneous methods{Heco [15], GraphMAE [27], HeCo++ [28]}.

Parameter Settings. For the baseline algorithms, we carefully select the parameters for each algorithm, following the procedures in the original papers. For all methods, we set the embedding dimension as 64.

We train our model for 1000 iterations using the Adam optimizer with the learning rate of 0.008 and set patience = 5. We uniformly set $\lambda = 0.5$ for all the datasets. The number of MultiHeadAttention heads is set to 8. We run the K-means algorithm 50 times to get an average value for all embedding learning baseline methods for fair comparison. The optimal values of A and B on ACM, Aminer, DBLP, and FreeBase are 4, 15, 1000, and 80, respectively; (7, 1) (3, 8), (6, 1) (18, 2). After conducting an analysis using the method of controlling variables, the optimal value for the number of neighbor samples was determined. We performed grid search to analyze the hyperparameters by training and evaluating each combination. Finally, we selected the hyperparameter combination that performed the best on the validation set as the final set of hyperparameters. The community detection results were evaluated using the ground-truth labels, and we reported the performance using two commonly adopted metrics, NMI and ARI in Table 1.

Table 1. Experimental Results on Datasets

Datasets	ACM		Aminer		DBLP		Freebase	
Models	NMI	ARI	NMI	ARI	NMI	ARI	NMI	ARI
GraphSage	29.20	27.72	15.74	10.10	51.50	36.40	9.05	10.49
GAE	27.42	24.49	28.58	20.90	72.59	77.31	19.03	14.10
Metapath2vec	48.43	34.65	30.80	25.26	73.55	77.70	16.47	17.32
HERec	47.54	35.67	27.82	20.16	70.21	73.99	19.76	19.36
HetGNN	41.53	34.81	21.46	26.60	69.79	75.34	12.25	15.01
HAN	48.44	47.19	31.86	32.73	73.42	78.75	19.62	19.72
HGT	53.29	49.82	31.92	32.81	73.85	79.37	19.64	19.81
DGI	51.73	41.16	22.06	15.93	59.23	61.85	18.34	11.29
DMGI	51.66	46.64	19.24	20.09	70.06	75.46	16.98	16.91
HeCo	56.87	56.94	32.26	28.64	74.51	80.17	20.38	20.98
HeCo++	60.82	60.09	**38.07**	**36.44**	75.39	81.20	20.62	21.88
GraphMAE	47.03	46.48	17.98	21.52	65.86	69.75	19.43	20.05
CDBMA	**64.17**	**68.78**	35.59	33.19	**76.43**	**81.34**	**22.51**	**21.10**

Experiment Result. The experimental results of different methods on different datasets are summarized in Table 1, where the values marked in bold are the best among all methods. Our method outperforms the comparison methods in evaluation metrics, with a significant improvement of approximately 10% in NMI on the ACM and Freebase datasets, and approximately 20% improvement in ARI on the ACM dataset. This demonstrates the effectiveness of our algorithm in community detection tasks. Metapath2vec performs random walks on meta-paths and outperforms GraphSage and GAE methods, which only consider structural features. HAN utilizes meta-paths and attention mechanisms to mine complex semantic information, leading to better results. Heterogeneous methods like Metapath2vec, HERec, HetGNN, HAN, DMGI, HeCo, and CDBMA outperform homogeneous methods such as GraphSage and GAE in capturing heterogeneous features and achieving better performance in community detection. Semi-supervised methods like HAN and HGT perform slightly better than some unsupervised heterogeneous methods by utilizing node labels in model training. HeCo++ utilizes hierarchical contrastive learning and shows superior performance over our proposed model on the Aminer dataset, but CDBMA demonstrates a more robust performance and surpasses HeCo++ on the majority of datasets.

Ablation Study. In this section, in order to analyze the effectiveness of different modules in our model, we conduct ablation experiments, The experimental results are summarized in Table 2. The comparison experiments are described as follows.-node: Remove the node attention mechanism in the semantic information encoder of the model.-type: Remove the type attention mechanism in the structural information encoder of the model.-semantic: Remove the semantic attention mechanism in the semantic information encoder of the model.-CDBMA: Represents the complete model without removing any module.

Table 2. The results of ablation experiment

Datasets	ACM		Aminer		DBLP		Freebase	
Models	NMI	ARI	NMI	ARI	NMI	ARI	NMI	ARI
-node	55.05	56.58	35.54	31.37	73.10	78.42	19.74	18.37
-type	41.46	36.52	33.65	34.66	**76.43**	**81.34**	18.11	13.68
-semantic	63.47	67.73	33.48	30.73	71.83	76.51	16.78	17.52
CDBMA	**64.17**	**68.78**	**35.59**	**33.19**	**76.43**	**81.34**	**22.51**	**21.10**

After removing the node attention mechanism from the semantic information encoder in the model, the overall performance decreased, indicating that the importance of each meta-path-based neighbor node varies for the target node. Removing the different type node aggregation layers from the structural information encoder led to a more significant decrease in overall performance, except for the DBLP dataset where the performance remained unchanged. This difference can be attributed to the DBLP dataset having only one type of neighbor nodes (papers), while the ACM dataset experienced a significant decrease. Removing the semantic attention mechanism from the semantic information encoder also resulted in a decrease in overall performance. The decrease was minor for the ACM dataset but significant for the DBLP and Freebase datasets. Based on these results, every component of the model is important for its performance.

4.5   Conclusion and Further Work

In this paper, we propose a community detection model for heterogeneous networks based on multi-attention mechanisms. Our model incorporates type attention, node attention, and semantic attention to capture the influence of different node types, the importance of high-order neighbor nodes, and the weights of meta-paths, respectively. We evaluate our model on four real-world datasets and demonstrate its superior performance compared to state-of-the-art algorithms, achieving approximately 10% improvement in NMI on the ACM and Freebase datasets, and approximately 20% improvement in ARI on the ACM dataset. In future work, we plan to explore task-driven or task-guided approaches to enhance the model's expressive power.

Acknowledgment. This work is funded by Key-Area Research and Development Program of Guangdong Province, China(2021B0101190002)

References

1. Chen, Z., Li, L., Bruna, J.: Supervised community detection with line graph neural networks. In: Proceedings of the International Conference on Learning Representations, LNCS, vol. 116, pp. 33–48. Springer, Cham (2019). https://doi.org/10.48550/arXiv.1705.08415

2. Hu, Z., Dong, Y., Wang, K., Sun, Y.: Heterogeneous graph transformer. In: Proceedings of the World Wide Web Conference (WWW). LNCS, vol. 12112, pp. 2704–2710. Springer, Cham (2020). https://doi.org/10.1007/978-3-030-45442-5_167

3. Sun, Y., Han, J., Yan, X., Yu, P.S., Wu, T.: Heterogeneous information networks: the past, the present, and the future. Proc. VLDB Endow **15**(12), 3807–3811 (2022). LNCS. Springer, Cham

4. Wang X., Ji, H., Shi, C., Wang, B., Ye, Y., Cui, P., Yu, P.S.: Heterogeneous graph attention network. In: Proceedings of the World Wide Web Conference (WWW). LNCS, vol. 11433, pp. 2022–2032. Springer, Cham (2019). https://doi.org/10.1145/3308558.3313417

5. Zhang, C., Song, D., Huang, C., Swami, A., Chawla, N.V.: Heterogeneous graph neural network. In: Proceedings of the ACM SIGKDD International Conference on Knowledge Discovery and Data Mining (KDD). LNCS, vol. 112, pp. 793–803. Springer, Cham (2019). https://doi.org/10.1145/3292500.3330961

6. Fu, X., Zhang, J., Meng, Z., King, I.: MAGNN: metapath aggregated graph neural network for heterogeneous graph embedding. In: Proceedings of the World Wide Web Conference (WWW). LNCS, vol. 12112, pp. 2331–2341. Springer, Cham (2020). https://doi.org/10.1145/3442381.3450080

7. Fang, Y., Yang, Y., Zhang, W., Lin, X., Cao, X.: Effective and efficient community search over large heterogeneous information networks. Proc. VLDB Endow. **13**(6), 854–867 (2020). LNCS. Springer, Cham

8. Luo, L., Fang, Y., Cao, X., Zhang, X., Zhang, W.: Detecting communities from heterogeneous graphs: a context path-based graph neural network model. In: Proceedings of the ACM International Conference on Information and Knowledge Management (CIKM). LNCS, vol. 2, pp. 1170–1180. Springer, Cham (2021). https://doi.org/10.1145/3459637.3482165

9. Wang, Z., Yu, D., Li, Q., Shen, S., Yao, S.: SR-HGN: semantic-and relation-aware heterogeneous graph neural network. Expert Syst. Appl. **224**, 119982 (2023). Springer, Cham. https://doi.org/10.1016/j.eswa.2021.119982

10. Wu, Y., Fu, Y., Xu, J., Yin, H., Zhou, Q., Liu, D.: Heterogeneous question answering community detection based on graph neural network. Inf. Sci. **621**, 652–671 (2023). Springer, Cham. https://doi.org/10.1016/j.ins.2021.04.037

11. Yue, S., Zhao, Y.: Research on overlapping community detection based on density peak clustering in heterogeneous networks. In: Proceedings of the International Conference on Intelligent Computing and Signal Processing (ICICSP), pp. 558–564. Springer, Cham (2022). https://doi.org/10.1007/978-3-030-92095-7_64

12. Zhao, J., Wang, X., Shi, C., Liu, Z., Ye, Y.: Network schema preserving heterogeneous information network embedding. In: Proceedings of the International Joint Conference on Artificial Intelligence (IJCAI), pp. 1366–1372. International Joint Conferences on Artificial Intelligence Organization (2020)

13. Hu, B., Fang, Y., Shi, C.: Adversarial learning on heterogeneous information networks. In: Proceedings of the ACM SIGKDD International Conference on Knowledge Discovery and Data Mining (KDD), pp. 120–129. Association for Computing Machinery (2019). https://doi.org/10.1145/3292500.3330851

14. Li, X., Ding, D., Kao, B., Sun, Y., Mamoulis, N.: Leveraging meta-path contexts for classification in heterogeneous information networks. In: Proceedings of the IEEE International Conference on Data Engineering (ICDE), pp. 912–923. IEEE (2021). https://doi.org/10.1109/ICDE51399.2021.00083

15. Wang, X., Liu, N., Han, H., Shi, C.: Self-supervised heterogeneous graph neural network with co-contrastive learning. In: Proceedings of the ACM SIGKDD Conference on Knowledge Discovery and Data Mining, pp. 1726–1736. Association for Computing Machinery (2021). https://doi.org/10.1145/3447548.3467268

16. Hamilton, W., Ying, Z., Leskovec, J.: Inductive representation learning on large graphs. In: Proceedings of the International Conference on Neural Information Processing Systems, pp. 1025–1035. Neural Information Processing Systems Foundation (2017)

17. Kipf, T.N., Welling, M.: Variational graph auto-encoders. stat **1050**, 21. arXiv (2016). https://doi.org/10.1145/3292500.3330851

18. Velickovic, P., Fedus, W., Hamilton, W.L., Li'o, P., Bengio, Y., Hjelm, R.D.: Deep graph infomax. In: Proceedings of the International Conference on Learning Representations. arXiv (2019)

19. Dong, Y., Chawla, N.V., Swami, A.: metapath2vec: Scalable representation learning for heterogeneous networks. In: Proceedings of the ACM SIGKDD International Conference on Knowledge Discovery and Data Mining, pp. 135–144. Association for Computing Machinery (2017). https://doi.org/10.1145/3097983.3098036

20. Shi, C., Hu, B., Zhao, W.X., Yu, P.: Heterogeneous information network embedding for recommendation. IEEE Trans. Knowl. Data Eng. **31**(2), 357–370 (2018). https://doi.org/10.1109/TKDE.2018.2868402

21. Park, C., Kim, D., Han, J., Yu, H.: Unsupervised attributed multiplex network embedding. In: Proceedings of the AAAI Conference on Artificial Intelligence, pp. 5371–5378. Association for the Advancement of Artificial Intelligence (2020)

22. Hu, Z., Dong, Y., Wang, K., Sun, Y.: Heterogeneous graph transformer. In: Proceedings of the Web Conference, pp. 2704–2710. Association for Computing Machinery (2020). https://doi.org/10.1145/3366423.3380198

23. Shi, Z., Wang, H., Leung, C.-S., Sum, J.: Constrained center loss for image classification. In: Yang, H., Pasupa, K., Leung, A.C.-S., Kwok, J.T., Chan, J.H., King, I. (eds.) ICONIP 2020. CCIS, vol. 1333, pp. 70–78. Springer, Cham (2020). https://doi.org/10.1007/978-3-030-63823-8_9

24. Satuluri, V., et al.: Simclusters: community-based representations for heterogeneous recommendations at Twitter. In: Proceedings of the ACM SIGKDD International Conference on Knowledge Discovery and Data Mining, pp. 3183–3193. Association for Computing Machinery (2020). https://doi.org/10.1145/3394486.3403203

25. Tommasel, A., Godoy, D.: Multi-view community detection with heterogeneous information from social media data. Neurocomputing **289**, 195–219 (2018). https://doi.org/10.1016/j.neucom.2018.01.053

26. Vaswani, A., et al.: Attention is all you need. In: Proceedings of the International Conference on Neural Information Processing Systems. Curran Associates Inc. (2017)

27. Hou, Z., et al.: GraphMAE: self-supervised masked graph autoencoders. In: Proceedings of the 28th ACM SIGKDD Conference on Knowledge Discovery and Data Mining, pp. 594–604 (2022)

28. Liu, N., Wang, X., Han, H., Shi, C.: Hierarchical contrastive learning enhanced heterogeneous graph neural network. IEEE Trans. Knowl. Data Eng. (2023)

An Adaptive Denoising Recommendation Algorithm for Causal Separation Bias

Qiuling Zhang, Huayang Xu, and Jianfang Wang[(✉)]

School of Computer Science and Technology, Henan Polytechnic University,
Jiaozuo 454000, China
wangjianfang@hpu.edu.cn

Abstract. In recommender systems, user selection bias often influences user-item interactions, e.g., users are more likely to rate their previously preferred or popular items. Existing methods can leverage the impact of selection bias in user ratings on the evaluation and optimization of recommendation system. However, these methods either inevitably contain a large amount of noise in the sampling process or suffer from the confound between users' conformity and interests. Inspired by the recent success of causal inference, in this work we propose a novel method to separate popularity biases for recommendation, named adaptive denoising and causal inference algorithm (ADA). We first compute the average rating of all feedback items of each user as the basis in converting explicit feedback to implicit feedback, and then obtain the true positive implicit data through adaptive denoising method. In addition, we separate the confounding of users' conformity and interest in the selection bias by causal inference. Specifically, we construct a multi-task learning model with regularization loss functions. Experimental results on the two datasets demonstrate the superiority of our ADA model over state-of-the-art methods in recommendation accuracy.

Keywords: Causal inference · Recommendation · Denoising · Regularization

1 Introduction

Recommender systems play an increasingly important role in solving the problems of information overload and users' non-specific needs in online applications such as e-commerce, news portals, and advertising [1–3], which identify users' potential preferences from observed interaction data to satisfy their personal interests. Traditional item recommendation predicts the rating of items that the target users do not participate in based on the historical information of neighboring users, and then generate a recommendation list. Depending on the type of feedback, recommender systems can be divided into explicit and implicit feedback. Unlike explicit feedback, implicit feedback (clicks, likes, or comments, among others) is more extensive and easier to collect. Even worse, most users' feedback is implicit. Therefore, it is important to consider how explicit feedback can be transformed into implicit feedback. The classical collaborative ranking methods, such as pointwise, pairwise and listwise, have achieved better practical results, especially the pairwise training model-BPR [4].

© The Author(s), under exclusive license to Springer Nature Singapore Pte Ltd. 2024
F. Wu et al. (Eds.): SMP 2023, CCIS 1945, pp. 188–201, 2024.
https://doi.org/10.1007/978-981-99-7596-9_14

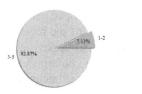

(a) Distribution of the mean value of the ratings   (b) Distribution of ratings of all users

Fig. 1. Netflix Prize user rating statistics

Most methods use uniform rule from explicit feedback to implicit, such as treating rating data of 5 as positive feedback and less than 5 as negative feedback [5]. However, this is not the case, in the Netflix Prize dataset contains more than 100 million ratings data from 480189 users for 17000 movies, there are significant differences in the habits and preferences of different users for rating as shown in Fig. 1(a). For example, about 92.8% of the mean value of the ratings above 3, while less than 7.2% rating between 1 and 2. Figure 1(b) shows that the proportion of all movies with a rating of 5 in the dataset is only 23.08%. Considering a rating of 5 as positive feedback only would result in a large amount of missing valid information, which may lead to incomplete separation of user interest and conformity.

During the conversion of explicit feedback into implicit feedback, four different types of feedback results will be generated: true positive, false positive, true negative, and false negative. False positives and false negatives are unavoidably noisy [5], while false negative interactions are much less common than true positive interactions in recommender systems, and false-positive interactions can have a worse impact on recommendation training. In this work, we focus on denoising false positive interactions. To keep the number of positive and negative samples consistent, we use randomly selected un-interacted data as negative samples. In the early stages of training, the loss value of the false positive interaction is greater than the loss value of the true positive interaction [6]. We determined the type of positive and negative feedback based on the average value of each user's ratings for all rated items. In the process of obtaining feedback types, our work can adaptively prune the interaction noise during training to accurately identify false positive interactions and true positive interaction information.

In implicit feedback, training data are usually taken from the user-item interaction logs as positive samples and exposure of no interaction behavior as negative samples. The interaction logs depend on which items are exposed, thus there are many biases [7]. Since users tend to focus on items that interest them and rarely pay attention to items that do not interest them, the observed items are not a representative sample of all items, resulting in the missing not at random, that is, a selection bias arises. There are typically confounding in selection bias: conformity (e.g., popular, or well-selling) and items of historical interest to users are displayed relatively more. The conformity describes the extent to which users are influenced by the results of others' recommendations. However, the interaction between users and items is also driven by user interest in most cases, the conformity does not necessarily truly reflect users' real interests.

Since user behavior data is obtained from observations rather than experiments in most cases, there is inevitably a user selection bias for items. Taking the model directly to fit the data and ignoring the bias will lead to poor performance, and to some extent, it also damages the user experience and fairness of the recommendation system. To separate the confounding factors of conformity and interest in user selection bias, we propose a regularization model that separates user conformity and interest in selection bias by causal inference, named denoising and causal inference (ADA), to remove the entanglement of selection bias, and construct a loss function with regularization, which makes the model more robust.

All in all, the main contributions of this work are as follows:

- We propose an ADA based on user ratings feedback and popularity biases, which effectively addresses a large amount of noise in the sampling process and separates the confound between users' conformity and real interests.
- We design a denoising model that adaptively classifies feedback types based on user ratings and then removes false positive noise.
- We design a separating popularity biases model with regularization loss functions which can separates user conformity and interest in selection bias by causal inference.
- We conduct extensive experiments on two public datasets to verify the effectiveness of the ADA on the state-of-the-art baseline.

2   Related Work

2.1   Denoising

Previous works typically collect various feedback from users (e.g., item properties [9], dwell time [8] and gaze patterns [10]) to distinguish users from implicit feedback generated by items based on user satisfaction and achieve noise removal purpose. To mitigate the effect of false positive interactions, previous approaches [11–13] have also considered including more feedback information in the training based on implicit feedback that could better reflect users' satisfaction with the items. However, these methods can only reduce the proportion of data noise without being able to eliminate much false information in the implicit feedback.

Some researchers propose to remove false positive data by developing an optimization function in converting explicit feedback to implicit feedback [13]. Wang et al. [6] address a new training strategy, named adaptive denoising training (ADT), which adaptively removes noise interactions during training. However, this method neglects how to classify the explicit feedback into implicit feedback more accurately.

2.2   Debias Recommendation

Recommender systems obtain data on user behavior based on observations rather than experiments. There are several biases, and user selection bias is a common bias in recommender systems [7, 14]. Previous works have shown that (1) recommendation models based on inverse propensity score estimation (IPS) can effectively mitigate the exposure bias of implicit feedback data [15]. (2) causal inference removes various biases in recommender systems by algorithmically revealing relevant data. Zheng et al. [5] propose an approach to model popularity bias from the perspective of causal analysis,

Guo et al. [16] mitigate the bias in the recommender system by considering social relationships while untangling. Zhang et al. [14] improve the model by removing the pernicious popularity bias. However, due to sparse data, the above models are relatively computationally intensive and tend to disappear the gradient, leading to information loss and overfitting. Therefore, to the best of our knowledge, there are very few methods that can effectively separate actual compliance and user interest by removing noise from the datasets.

3 Methodology

3.1 Implicit Feedback Based on Average

To obtain a more precise type of implicit feedback, we obtain positive and negative feedback by calculating the mean value for all items rated by each user. Those greater than the mean are positive feedback and those less than the mean are negative feedback. That is to say, the rating $r_{u,i}$ of each user u to item i, the average of its ratings is $\overline{r_u}$, if $r_{u,i} \geq \overline{r_u}$, the user-item rating is 1, otherwise, the one is 0.

3.2 Adaptive Denoising

Without any side data, Wang et al. [6] use a truncated cross-entropy (T_CE) loss that simply uses the loss values generated by user-item interactions on the input model, and discard positive interactions with larger cross-entropy loss value:

$$L_{T\_CE}(u, i) = \begin{cases} 0, & L_{T\_CE}(u, i) > \tau \wedge \bar{y}_{ui} = 1 \\ L_{T\_CE}(u, i), & otherwise \end{cases} \tag{1}$$

Where τ is a predefined threshold. $\bar{y}_{u,i} = 1$ indicates that user u and item i had a positive interaction. The T-CE loss removes any positive interactions with CE loss larger than τ from the training. Although this simple T_CE loss is easy to interpret and implement, however, the loss value decreases with increasing training iterations during the removal of false positives. Since the loss values vary for different data sets, we replace the fixed threshold with a dynamic threshold and design a dynamic function $\varepsilon(T)$ instead of the dynamic threshold τ for T-CE loss, the dynamic function $\varepsilon(T)$ is given as follow:

$$\varepsilon(T) = \min(\alpha T, \varepsilon_{\max}) \tag{2}$$

where $\varepsilon_{\max}$ is the upper limit and α is the super parameter to adjust the speed to achieve the maximum rate of decline.

3.3 Causal Separation Conformity and Interest

There are two main points in user selection bias: (1) the popularity of item to users, and (2) the user interest in the item's characteristics. Inspired by the literature [5]. If a user interacts with an item with low popularity, but no interaction with an item with high popularity, we infer that the user likes the item, separate the conformity and interest on interaction (click) and construct a loss function to obtain a top-K recommendation list for the items. We propose further optimization of the loss function with L2 regularization.

1) Separation of conformity and interest.

Case1: Assuming that user u has interaction behavior for an item a with high popularity and no interaction behavior for an item b with low popularity.

$$M_{ua}^C > M_{ub}^C$$
$$M_{ua}^I + M_{ua}^C > M_{ub}^I + M_{ub}^C \tag{3}$$

where M^I is the interest matrix and M^C is the conformity influence matrix.

Case2: User u interacts with a low-popularity item c and no interaction with a high-popularity item d. Then the user must be interested in item c.

$$M_{uc}^I > M_{ud}^I, M_{uc}^C < M_{ud}^C$$
$$M_{uc}^I + M_{uc}^C > M_{ud}^I + M_{ud}^C \tag{4}$$

By separating the conformity and interest preferences, and modeling each part using BPR loss, the loss of each part can be expressed as follows:

$$L_{con}^{O_1} = \sum_{(u,i,j) \ni O_1} BPR\left(< u^{con}, i^{con} >, < u^{con}, j^{con} >\right)$$
$$L_{con}^{O_2} = \sum_{(u,i,j) \ni O_1} -BPR\left(< u^{con}, i^{con} >, < u^{con}, j^{con} >\right)$$
$$L_{int}^{O_2} = \sum_{(u,i,j) \ni O_2} BPR\left(< u^{int}, i^{int} >, < u^{int}, j^{int} >\right) \tag{5}$$
$$L_{int}^{O_1+O_2} = \sum_{(u,i,j) \ni O} BPR\left(< u^t, i^t >, < u^t, j^t >\right)$$

where u^t, i^t, j^t is interest and two kinds of embedding after aggregation of conformity respectively, O is the overall training sample, O_1 is the sample of case1 and O_2 is the sample of case2:

$$u^t = u^{int}||u^{con}, i^t = i^{int}||i^{con}, j^t = j^{int}||j^{con} \tag{6}$$

where $||$ is the aggregation function, using concat method aggregation.

2) Constructing the loss function.

We construct a multi-task learning model with a mixture of supervised and unsupervised loss functions after obtaining the reasons for the selection bias:

$$L(\theta) = L_{click}^{O_1+O_2}(\theta) + \alpha(L_{interest}^{O_2}(\theta) + L_{conformity}^{O_1+O_2}(\theta)) + \beta L_{discrepancy}(\theta) + \lambda \|\theta\|^2 \tag{7}$$

where α, β, λ is the hyper-parameter, θ is the parameter of the model, L_{click} is the estimated number of clicks loss function, $L_{interest}$ is the interest modeling loss function, $L_{conformity}$ is the conformity modeling loss function. We refer to [5] for details on $L_{discrepancy}$. O_1 and O_2 is negative samples that are not as popular as positive samples and negative samples that are more popular than positive samples, respectively.

4 Experiments

In this section, we aim to answer the following research questions:

- RQ1: How does our proposed removal false positive feedback with state-of-the-art methods? In addition is it necessary to use the conversion rule?
- RQ2: What is the impact of separating conformity and interest by causal inference on the effectiveness of ADA?
- RQ3: Does loss function with regularization achieve the anti-overfitting?

4.1 Experimental Setting and Datasets

For the IPS-based model, we fix the size of the embedding to 128. While for CausE, DICE and ADA, the size of the embedding is fixed to 64 as they contain two sets of embeddings. Therefore, the number of parameters is the same for all methods to ensure a fair comparison. We set $\alpha = 0.1$, $\beta = 0.01$, and $\lambda = 0.0001$, as this shows better performance and has no effect on either the dataset or the backbone model. We use BPR [4] as the loss function for all baselines and use Adam [24] for optimizing all parameters with the learning rate initialized to 0.001. The other hyperparameters of our method and baseline are tuned by grid search.

Experiments were conducted on two publicly accessible MovieLens (ML)-1M [17] and Netflix Prize [18] datasets. Interaction data from 5% of the users are randomly selected in the Netflix Prize for data processing.

4.2 Baselines

Causal methods usually serve as additional models upon backbone recommendation frameworks [5]. We choose the classical methods IPS [19, 20], CausE [21], and DICE [5] for applying causal inference to remove model bias in recommender systems. Meanwhile, we also compare the performance by the above three methods on the MF [4] and GCN [22]. Specifically, we use BPR-MF [4] and LightGCN [23], which are both state-of-the-art recommendation models.

4.3 Evaluation

We follow the full-ranking protocol by ranking all the non-interacted items for each user [7]. For performance comparison, we adopt three widely used metrics Recall@K, HR@K and NDCG@K over the top-K items, where K is set as 20 or 50.

4.4 Data Preprocessing

The original data are preprocessed by 5 different data processing methods such as 1) ONLY 5, 2) Average (AVE), 3) Truncated cross-entropy (T_CE) denoising, 4) Random (RAN) denoising and 5) AVE with the same user and item number of the T_CE.

1) ONLY 5: We binarize the datasets by keeping ratings of five stars as one, and others as zero [5].

2) AVE: We divide the data into positive and negative samples based on the mean value of each user's rating. To ensure data quality and to give the denoising algorithm sufficient space to truncate the data, we use a 20-core threshold, i.e., discard users and items with less than 20 positive interactions.

3) T_CE denoising: According to the data obtained from Experiment 2), we use the adaptive denoising method to process the data. In GMF, the training batch epoch is set to 10 times, the data batch size batch size is 1024, and the hyperparameter εmax = 0.1 in view of the low sparsity of the ML-1M. While the sparsity of the Netflix Prize is high, the hyperparameter $\varepsilon_{max} = 0.05$. We select the remaining data from the last batch of training as positive samples.

4) RAN denoising: We record the number of truncated positive data in T_CE and randomly discard positive data based on AVE dataset, However, it is necessary to ensure that the number of user-item interactions is equal to the number of T_CE.

5) AVE_T_CE denoising: In view of the fact that the noise reduction itself inevitably causes all interactions for some elements to be truncated, resulting in the elimination of a small number of items from the dataset. We eliminate only those items that are completely removed by the T_CE algorithm based on AVE, the number of users and items of AVE_T_CE method is the same as that of T_CE. The only difference is that T_CE still truncates the internal interactions with existing user entries. However, the internal interaction of AVE_T_CE method has not changed. Our aim is to exclude the influence of the reduced number of items due to the denoising of the data by T_CE on the experimental results.

To measure the performance of causal learning in a non-IID environment, an intervention test set is required. Therefore, all datasets are transformed according to a standard protocol [5, 25]. We use the training set, intervention training set, validation set, and test set in 6:1:1:2 ratio.

Table 1 shows the statistical information after data processing by the five methods, ONLY-5 means that 5 ratings in explicit feedback are regarded as positive feedback and others are negative feedback [5].

Table 1. Dataset pre-processed statistics

Dataset		User	Item	Interaction	Ent Train	Ent Test
ML-1M	ONLY_5	4627	1865	212247	6.13	7.35
	AVE	4797	2342	517520	6.62	7.61
	T_CE	4797	2287	480421	6.49	7.47
	RAN	4797	2342	480421	6.62	7.61
	AVE_T_CE	4797	2287	511413	6.60	7.59
Netflix Prize	ONLY_5	16160	6381	1071768	6.77	8.41
	AVE	17955	7398	2586700	7.04	8.56

(continued)

Table 1. (*continued*)

Dataset		User	Item	Interaction	Ent Train	Ent Test
	T_CE	17955	6227	2323988	6.86	8.24
	RAN	17955	7398	2323988	7.04	8.56
	AVE_T_CE	17955	6227	2547460	7.02	8.44

Ent indicates the entropy value of the number of interactions of all items.

$$O_{entropy} = \sum_{i \in I} (-\frac{i_{count}}{\sum\limits_{i \in I} i_{count}} \log(\frac{i_{count}}{\sum\limits_{i \in I} i_{count}})) \tag{8}$$

where i_{count} is the total times of all user interactions for (i).

4.5 Performance of Denoising (RQ1)

To demonstrate the denoising effect of our recommendation model, the experimental results of top-K = 20 and 50 recommendations are shown in Table 2 and Table 3, which summarizes the recommendation performance comparison of MF and GCN model with ONLY_5, AVE, T_CE, RAN, AVE_T_CE over two datasets.

Table 2. Recommendation performance in different methods on ML-1M

Dataset		ML-1M					
		top-K = 20			top-K = 50		
Model	Data Processing	Recall	HR	NDCG	Recall	HR	NDCG
MF	ONLY_5	0.10515	0.47633	0.07631	0.19703	0.67624	0.10693
	AVE	0.11668	0.73983	0.12496	0.21152	0.87825	0.15345
	T_CE	**0.12407**	**0.74275**	**0.13050**	**0.22892**	**0.88680**	**0.16314**
	RAN	0.09864	0.66008	0.10013	0.18708	0.83010	0.12891
	AVE_T_CE	0.11742	0.73274	0.12423	0.21025	0.87242	0.15203
GCN	ONLY_5	0.12777	0.53706	0.09630	0.24142	0.72120	0.13421
	AVE	0.13531	0.76839	0.14685	0.24265	0.88576	0.17905
	T_CE	**0.14821**	**0.78736**	**0.15738**	**0.26991**	**0.90431**	**0.19524**
	RAN	0.12362	0.72983	0.12419	0.23085	0.87658	0.15900
	AVE_T_CE	0.13618	0.76547	0.14586	0.24777	0.89639	0.18008

Table 3. Recommendation performance in different methods on Netflix

Dataset		Netflix					
		top-K = 20			top-K = 50		
Model	Data Processing	Recall	HR	NDCG	Recall	HR	NDCG
MF	ONLY_5	0.10306	0.49510	0.08523	0.17611	0.65084	0.10742
	AVE	0.10217	0.68922	0.12468	0.17735	0.82222	0.14114
	T_CE	**0.11814**	**0.72080**	**0.13725**	**0.20990**	**0.84700**	**0.16018**
	RAN	0.09295	0.63263	0.09980	0.16549	0.78652	0.11935
	AVE_T_CE	0.10621	0.69757	0.12917	0.18529	0.83057	0.14654
GCN	ONLY_5	0.11312	0.52363	0.09452	0.19665	0.67969	0.12043
	AVE	0.09940	0.68075	0.12465	0.17641	0.81687	0.14153
	T_CE	**0.12359**	**0.72932**	**0.14552**	**0.22149**	**0.85808**	**0.16991**
	RAN	0.08899	0.62205	0.09666	0.16445	0.78407	0.11748
	AVE_T_CE	0.10693	0.70108	0.13234	0.19404	0.84266	0.15299

We have the following observations:

- We note that the method T_CE for the models MF and GCN shows excellent performance in all metrics. This indicates that method T_CE can effectively reduce the interference from false-positive interactions, reducing the data noise in the training input model and thus making the final model more robust.
- The random method has the worst performance because it removes the data randomly and most of the removed data are true-positive interactions because there are fewer false-positive interactions in the data itself, which hinders the construction of the model and leads to worse metrics.
- The lower performance of AVE_T_CE than T_CE indicates that the T_CE denoising method is effective in removing false-positive noise with the same number of users and items.

4.6 Effect of Causality on Performance (RQ2)

To verify the effect of separating the conformity and interest, we use classical frameworks (CausE, IPS and DICE) to compare with our proposed ADA, which are based on the models MF and GCN. For the traditional causal recommendation DICE model, which is no denoising operation. Instead, ADA uses the T_CE algorithm to denoise the data. Table 4 and Table 5 show the results on ML-1M and Netflix, respectively. "RI" denotes the Relative Improvement of ADA over baselines on average.

For the recommendation task, ADA consistently outperforms all the baselines on all the evaluation metrics. From the reported results, we have the following observations:

Table 4. Recommendation performance in denoising of different methods on ML-1M

Dataset		ML-1M						RI
		top-K = 20			top-K = 50			
Model	Methods	Recall	HR	NDCG	Recall	HR	NDCG	
MF	CausE	0.12239	0.73733	0.13286	0.21706	0.87596	0.16016	21.01%
	IPS	0.12338	0.74296	0.12541	0.22609	0.88492	0.15862	21.04%
	DICE	0.13949	0.77611	0.14525	0.25833	0.90514	0.18254	9.00%
	ADA	**0.15549**	**0.80341**	**0.16149**	**0.29222**	**0.92328**	**0.20577**	-
GCN	CausE	0.12042	0.73462	0.13400	0.21901	0.86845	0.16229	31.78%
	IPS	0.14445	0.78215	0.15057	0.26467	0.90681	0.18940	15.77%
	DICE	0.15690	0.80842	0.16696	0.28455	0.91911	0.20651	8.29%
	ADA	**0.17535**	**0.83781**	**0.18383**	**0.31840**	**0.93495**	**0.22847**	-

Table 5. Recommendation performance in denoising of different methods on Netflix

Dataset		Netflix						RI
		top-K = 20			top-K = 50			
Model	Methods	Recall	HR	NDCG	Recall	HR	NDCG	
MF	CausE	0.09101	0.64383	0.11406	0.16605	0.79754	0.13015	30.12%
	IPS	0.10843	0.70270	0.13058	0.19828	0.83959	0.15401	13.57%
	DICE	0.10693	0.70108	0.13234	0.19404	0.84266	0.15299	14.12%
	ADA	**0.12897**	**0.74369**	**0.15417**	**0.23324**	**0.87312**	**0.18013**	-
GCN	CausE	0.09557	0.65892	0.11929	0.17192	0.80289	0.13566	31.15%
	IPS	0.09491	0.66065	0.12704	0.17958	0.81069	0.14594	27.05%
	DICE	0.11905	0.73182	0.14909	0.21044	0.85742	0.16934	9.84%
	ADA	**0.13724**	**0.75817**	**0.16433**	**0.24271**	**0.88120**	**0.18933**	-

- The ADA achieves the best performance in all recommendation metrics. An important reason is the use of adaptive denoising in the AD, which can effectively prune the false positives in the interaction data and thus weaken the interference of false positives in the model training. It makes the positive sample selection of the causal model more reliable and can correct the popularity of the sample, thus efficiently separating the interest and popularity between users and items.

- The causal recommendation models achieve better performance in most cases because their framework uses separate embeddings to represent and separate conformity and interest in different scenarios.
- It is worth mentioning that in the Netflix Prize dataset in the MF model using IPS approach, it shows better performance in some metrics compared to DICE. The reason is that DICE requires higher quality data, and when false positive interactions exist in the interaction data used, the causal separation of the model is somewhat disturbed, which leads to suboptimal recommendation results.

4.7   Regularization Impact (RQ3)

We evaluate the influence of the hyper-parameter λ of three metrics by changing the L2 regularization coefficients in ML-1M. Trend of performance L_2 is shown in Fig. 2.

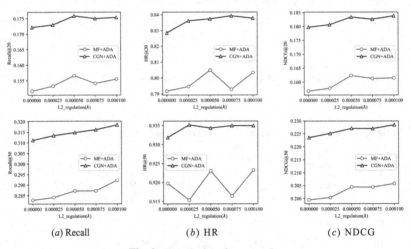

Fig. 2. Trend of performance L_2

We have three major observations:

- In Fig. 2(a) that the Recall increases with the increase of the regularization factor, as the model falls into overfitting after several training sessions, which leads to poor performance in the test set. By adding the L_2 regularization term, the overfitting of the model is limited, allowing the model to generalize better. At the same time, we can also see that when the L_2 regularization coefficient is larger, the metrics of the model decrease, such as the recall metric for top-K = 20 on the top, because the model becomes underfitted after optimizing the model parameters in the late stage of model training, which is affected by the L2 regularization coefficient.
- For HR in Fig. 2(b), the overall gain of the GCN + ADA model is smooth and stable. However, the MF + ADA fluctuates strongly, especially when top-K = 50. The reason is that the HR reaches overall saturation and the L_2 regularization term may adjust and affect the learning ability of the model. Therefore, the intervention for the model with more saturated training leads to large fluctuations.

- In Fig. 2(c) of NDCG, we can see that the indicators at top-K = 20 and top-K = 50 have an overall increasing trend and the overall metrics with the addition of L_2 regularization have achieved improvement with the increase of the regularization factor. This also proves that the model is more robust after adding the L_2 regularization factor.

We find that the model performs relatively best when the regularization factor λ = 0.0001. We used this coefficient to conduct the same comparison experiments on the Netflix Price, and we chose the DICE, ADA and ADA without using regularization coefficient method for the comparison experiments, and the comparison results are shown in Fig. 3.

As can be seen in Fig. 3, the ADA achieves the best performance in all recommendation metrics on the Netflix Prize as well. In contrast, the recommended metrics decreased in both the MF and GCN models after removing the L_2 regularization term, which effectively illustrates the necessity of adding the L_2 regularization term to the ADA. Meanwhile, even after removing the L_2 regularization term, the metrics of the ADA also outperform the DICE, which again proves the effectiveness of the denoising method in the ADA.

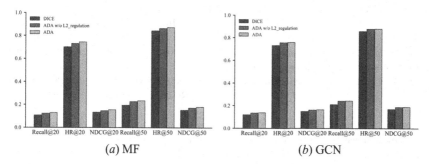

(a) MF   (b) GCN

Fig. 3. Comparison of different methods on Netflix Prize

5 Conclusions

In this work, we propose algorithms for causal inference that can remove false positive noise from implicit feedback and separate conformity from interest. First, we extract more accurate user-item information from observed interactions based on the average of utility item ratings. Second, for the noise in implicit feedback, we apply the adaptive threshold denoising function to obtain the true positive feedback data during user interaction without additional information. To improve the accuracy of recommendation prediction, we use a causal recommendation approach to separate the user's conformity and interest, and develop a causal loss function with regularization. Experiments show that our method is more robust in a non-IID environment with denoised data and has higher recommendation accuracy compared to other state-of-the-art algorithms.

References

1. Chen, C., et al.: An efficient adaptive transfer neural network for social-aware recommendation. In: Proceedings of the 42nd International ACM SIGIR Conference on Research and Development in Information Retrieval, vol. 225 (2019)
2. Lin, T., Gao, C., Li, Y.: Cross: Cross-platform recommendation for social e-commerce. In: Proceedings of the 42nd International ACM SIGIR Conference on Research and Development in Information Retrieval, vol. 515 (2019)
3. Lu, H., Zhang, M., Ma, S.: Between clicks and satisfaction: Study on multi-phase user preferences and satisfaction for online news reading. In: The 41st International ACM SIGIR Conference on Research & Development in Information Retrieval, vol. 435 (2018)
4. Steffen, R., Christoph, F., Zeno, G., Lars, S.: Bayesian personalized ranking from implicit feedback. In: Proceedings of the twenty-fifth conference on uncertainty in artificial intelligence. AUAI Press (2009)
5. Zheng, Y., Gao, C., Li, X., He, X., Li, Y., Jin, D.: Disentangling user interest and conformity for recommendation with causal embedding. In: Proceedings of the Web Conference 2021, vol. 2980 (2021)
6. Wang, W., Feng, F., He, X., Nie, L., Chua, T.: Denoising implicit feedback for recommendation. In: Proceedings of the 14th ACM international conference on web search and data mining, vol. 373 (2021)
7. Chen, J., Dong, H., Wang, X., Feng, F., Wang, M., He, X.: Bias and debias in recommender system: A survey and future directions. ACM Trans. Inf. Syst. **41**(3), 1–39 (2023)
8. Kim, Y., Hassan, A., White, R.W., Zitouni, I.: Modeling dwell time to predict click-level satisfaction. In: Proceedings of the 7th ACM International Conference on Web Search and Data Mining, vol. 193 (2014)
9. Lu, H., et al.: Effects of user negative experience in mobile news streaming. In: Proceedings of the 42nd International ACM SIGIR Conference on Research and Development in Information Retrieval, vol. 705 (2019)
10. Zhao, Q., Chang, S., Harper, F.M., Konstan, J.A.: Gaze prediction for recommender systems. In: Proceedings of the 10th ACM Conference on Recommender Systems, vol. 131 (2016)
11. Frolov, E., Oseledets, I.: Fifty shades of ratings: how to benefit from a negative feedback in top-N recommendations tasks. In: Proceedings of the 10th ACM Conference on Recommender Systems, vol. 91 (2016)
12. Lavee, G., Koenigstein, N., Barkan, O.: When actions speak louder than clicks: A combined model of purchase probability and long-term customer satisfaction. In: Proceedings of the 13th ACM Conference on Recommender Systems, vol. 287 (2019)
13. Liu, C., White, R.W., Dumais, S.: Understanding web browsing behaviors through Weibull analysis of dwell time. In: Proceedings of the 33rd International ACM SIGIR Conference on Research and Development in Information Retrieval, vol. 379 (2010)
14. Zhang, Y., et al.: Causal intervention for leveraging popularity bias in recommendation. In: Proceedings of the 44th International ACM SIGIR Conference on Research and Development in Information Retrieval, vol. 11 (2021)
15. Huang, J., Oosterhuis, H., de Rijke, M.: It is different when items are older: debiasing recommendations when selection bias and user preferences are dynamic. In: Proceedings of the Fifteenth ACM International Conference on Web Search and Data Mining, vol. 381 (2022)
16. Sheth, P., Ruocheng Guo, L., Cheng, H.L., Candan, K.S.: Causal disentanglement for implicit recommendations with network information. ACM Trans. Knowl. Discovery Data **17**(7), 1–18 (2023)
17. Maxwell Harper, F., Konstan, J.A.: The MovieLens datasets: History and context. ACM Trans. Interact. Intell. Syst. **5**(4), 1–19 (2015). https://doi.org/10.1145/2827872

18. Bennett, J., Lanning, S.: The netflix prize. In: Proceedings of KDD Cup and Workshop, vol. 2007, p. 35. New York (2007)
19. Joachims, T., Swaminathan, A., Schnabel, T.: Unbiased learning-to-rank with biased feedback. In: Proceedings of the Tenth ACM International Conference On Web Search and Data Mining, vol. 781 (2017)
20. Schnabel, T., Swaminathan, A., Singh, A., Chandak, N., Joachims, T.: Recommendations as treatments: Debiasing learning and evaluation. In: International Conference on Machine Learning, vol. 1670. PMLR (2016)
21. Bonner, S., Vasile, F.: Causal embeddings for recommendation. In: Proceedings of the 12th ACM Conference on Recommender Systems, vol. 104 (2018)
22. Kipf, T.N., Welling, M.: Semi-supervised classification with graph convolutional networks (2016). arXiv preprint arXiv:1609.02907
23. He, X., Deng, K., Wang, X., Li, Y., Zhang, Y., Wang, M.: LightGCN: simplifying and powering graph convolution network for recommendation. In: Proceedings of the 43rd International ACM SIGIR Conference on Research and Development in Information Retrieval (2020)
24. Kingma, D.P., Ba, J.: Adam: A method for stochastic optimization (2014). arXiv preprint arXiv:1412.6980
25. Zhu, Y., Ma, J., Li, J.: Causal Inference in Recommender Systems: A Survey of Strategies for Bias Mitigation, Explanation, and Generalization (2023). arXiv preprint arXiv:2301.00910

Tuning Query Reformulator
with Fine-Grained Relevance Feedback

Yuchen Zhai[1], Yong Jiang[2], Yue Zhang[2], Jianhui Ji[2], Rong Xiao[2],
Haihong Tang[2], Chen Li[2], Pengjun Xie[2], and Yin Zhang[1(⊠)]

[1] College of Computer Science and Technology, Zhejiang University, Hangzhou,
China
{zhaiyuchen,zhangyin98}@zju.edu.cn
[2] Alibaba Group, Hangzhou, China
{shiyu.zy,jianhui.jjh,xiaorong.xr,piaoxue,puji.lc,
chengchen.xpj}@alibaba-inc.com

Abstract. Pseudo-relevance feedback (PRF) has been empirically validated as an effective query reformulation method to improve retrieval performance. Recent studies formulate query reformulation as a reinforcement learning task to directly optimize the retrieval performance. However, this paradigm computes the feedback signals by comparing the retrieved documents with the manual annotations, and neglects that the annotations severely suffer from the unlabeled problem (the relevant documents of a query may not be fully annotated), causing the model to overfit the training set. Moreover, the training of reinforcement learning is expensive and unstable. To address the above problems, inspired by recent great achievements of reinforcement learning from human feedback (RLHF), we propose a simple fine-grained feedback framework for query reformulation, which computes the feedback signals by a powerful re-ranking model instead of manual annotations. Specially, we first utilize various automation methods to generate annotated data, which allows us to initialize the reformulator and obtain a good starting point. Then we employ a re-ranking model to assign fine-grained scores to the rewritten queries generated by the reformulator. Finally, we refine the reformulator using feedback scores. In this way, the knowledge of the re-ranking model can be effectively transferred to the reformulator, leading to a better generalization performance. Furthermore, our framework can enhance performance by leveraging a large amount of unlabeled data. Experiments on a real-world E-Commerce search engine and three public benchmarks demonstrate the effectiveness of our framework.

Keywords: query reformulation · sparse retrieval · pseudo-relevance feedback

1 Introduction

Users of the Internet typically formulate short and ambiguous queries due to the difficulty of expressing their information needs precisely in words. This difficulty, also known as a "lexical chasm", is caused by a vocabulary mismatch

F. Wu et al. (Eds.): SMP 2023, CCIS 1945, pp. 202–217, 2024.
https://doi.org/10.1007/978-981-99-7596-9_15

between user queries and web documents, as they may use different language styles and vocabularies. Decades of IR research demonstrate that such casual queries prevent a search engine from correctly and completely satisfying users information needs. An effective way to address the aforementioned problem is to automatically rewrite a query so that it becomes more likely to retrieve relevant documents. This technique is known as query reformulation. It often leverages external resources (such as Thesaurus [45] and Relevance Feedback [19]) to modify the query.

Pseudo Relevance Feedback (PRF) has been demonstrated to be one of the most effective methods for query reformulation in information retrieval [22,25, 28,40,47]. This approach assumes that the top-ranked documents returned by the retrieval system are relevant to the user's information needs. Terms obtained from the top-ranked documents can be incorporated into the original query to form a revised query, which is subsequently utilized to retrieve the final set of documents presented to the user. However, since most useful PRF terms are typically unknown, many existing methods make strong assumptions based on heuristic rules to estimate the importance of each PRF term. These assumptions are not necessarily satisfied with the ultimate objective of optimizing the retrieval performance in the information retrieval task.

Recently, some studies [32,35] have proposed a reinforcement learning framework for automatic query reformulation based on the PRF, which aims to optimize retrieval metrics directly. As shown in Fig. 1, they treat the reformulator as a policy network to revise an initial query. The retrieval model serves as an environment and produces the retrieved documents that make up the state. By comparing the retrieved documents to the manual annotations, we are able to compute the retrieval performance (i.e., feedback signals). However, the RL framework requires large-scale fine-grained annotations (labeling all relevant documents of a query) to generalize well. Since the number of relevant documents for a query is typically variable, and the documents are extracted from an extremely large collection, collecting the fine-grained complete annotations for information retrieval is challenging. In this way, some relevant documents are inevitably missed by human annotators. This severely affects the RL framework which directly treats these missing annotations as negative instances to compute the feedback signals. Moreover, reinforcement learning is historically unstable during training due to the sparse reward provided by the environment and the difficulty of learning complex tasks from scratch.

To address the above problems, inspired by recent great achievements of reinforcement learning from human feedback (RLHF) in large-scale pre-trained language models (LLMs) [3,9,27,36], we propose a novel **Fine-gRained feEDback** framework for query reformulation (**FRED**) that is able to eliminate the misguidance of missing annotations in training. Unlike previous methods that determine the coarse reward directly from the manual annotations, our method computes a fine-grained feedback signal by a powerful relevance re-ranking model. This avoids giving a confusing signal to the reformulator with unlabeled relevant documents. Specifically, we first use various automated strategies to generate a

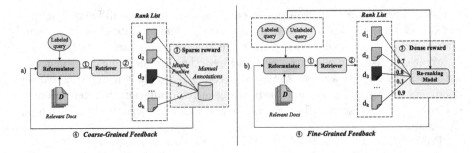

Fig. 1. Comparing our framework with previous reinforcement learning methods in the training phase. Our framework consists of a reformulator, a retriever, and an initialized powerful re-ranking model. 1) The reformulator first utilizes labeled query, unlabeled query, and relevant documents to generate multiple rewritten candidates. 2) These candidates are fed to the retriever to obtain the ranked documents. 3) The re-ranking model calculates a dense reward by evaluating the relevance between the original queries and the ranked documents. 4) The dense reward provides fine-grained feedback to refine the reformulator.

dataset for supervised query reformulation, which avoids training the model from scratch and enables fast convergence of the reformulator. Next, the initialized reformulator is used to regenerate the dataset. Then, similar to RLHF which trains a powerful reward model to refine the policy network, our framework fine-tunes a BERT-based re-ranking model based on a labeled dataset. We utilize the re-ranking model to compute fine-grained feedback signals for the dataset. Lastly, the feedback signals are utilized to further fine-tune the reformulator, which alleviates the missing label problem in the manual annotations. In addition, by utilizing weak supervision signals from the re-ranking model, our model can be further trained on unlabeled data to enhance its performance.

Our main contributions are as follows:

- We propose a novel query reformulator tuning framework to mitigate the confusing learning signals caused by the missing annotations, which bridges the gap between reinforcement learning and the query reformulation task.
- Aided by the relevance re-ranking model, we propose a semi-supervised method to further enhance the generalization of the proposed framework.
- We conduct a series of experiments on a real-world E-Commerce search engine and three public benchmarks. The results demonstrate the effectiveness of the proposed methods.

2   Methodology

In this section, we focus on the problem of query reformulation for information retrieval. Given an original query q, the goal is to reformulate q as a new query q', which is expected to retrieve more relevant documents.

2.1 Framework

Our framework consists of two stages: 1) the cold start stage and 2) the feedback adjustment stage. In the first stage, we use various automated methods to generate multiple rewritten queries $\hat{\mathbf{q}}_i = (\hat{q}_i^1, \ldots, \hat{q}_i^m)$ for each original query q_i. These rewritten queries can be used to initialize our reformulator. The reformulator is trained with a standard cross entropy (CE) loss $Loss_{CE}$ to maximize the likelihood of generating $\hat{\mathbf{q}}$. This stage enables our model to converge fast and have a good starting point. In the second stage, we utilize the refined labeled data and the unlabeled data (query without labeling corresponding relevant documents) to further fine-tune the reformulator. Specifically, for q_i, we first use the initialized reformulator in the first stage to sample candidate targets $\tilde{\mathbf{q}}_i = (\tilde{q}_{i1}, \ldots, \tilde{q}_{il})$. Next, the candidate target $\tilde{q}_{ij}$ is fed into a search engine to retrieve the relevant documents. Our framework feeds these documents and the original query q_i into a re-ranking model to compute fine-grained scores s_{ij} for $\tilde{q}_{ij}$. We use these scores to filter out candidates with low confidence and refine the candidate set to $\tilde{\mathbf{q}}_i = \{\tilde{q}_{ij} | s_{ij} > \omega \wedge \tilde{q}_{ij} \in \tilde{\mathbf{q}}_i\}$, where ω is a hyperparameter. Then the refined candidate set is viewed as the training target, and we get the probability of the reformulator in an auto-regressive manner.

$$P(\tilde{q}_{ij} \mid q_i) = \prod_{k=1}^{|\tilde{q}_{ij}|} P\left(\tilde{q}_{ij}^k \mid q_i, \tilde{q}_{ij}^{1:k-1}\right) \tag{1}$$

After that, the training loss for $\tilde{q}_i$ becomes:

$$Loss_{IF} = -\frac{1}{|\tilde{\mathbf{q}}_i|} \sum_{j=1}^{|\tilde{\mathbf{q}}_i|} s_{ij} \log P\left(\tilde{q}_{ij} \mid q_i\right) \tag{2}$$

Finally, the total loss on the $\hat{\mathbf{q}}$ and $\tilde{\mathbf{q}}$ can be defined as:

$$Loss = \alpha Loss_{IF} + (1 - \alpha)Loss_{CE} \tag{3}$$

where α is a hyper-parameter. In this way, we can obtain the training objectives for the unlabeled and labeled data.

During the testing phase, we generate multiple candidate queries for each query. These queries are further filtered by a re-ranking model, and the most appropriate candidate query is selected as the final rewritten result q'.

2.2 Reformulator

The reformulator comprises the following components.

Relevance Set. The relevance set $U = (U_0, U_1, \ldots, U_n)$ consists of the top retrieved feedback documents.

Input. The input to the reformulator is a concatenation of the original query and the relevance set, which is separated with a special token "[SEP]".

Generator. We employ the T5 [39] as the base model architecture and generate candidate queries by applying the beam search strategy [9].

Training Objective. During the cold start stage, the target queries are generated by a series of automated methods [7], e.g., Pseudo-relevance Feedback and Semantic Analysis. To further enhance the performance, we manually annotate a query rewriting dataset for the E-Commerce search engine. During the feedback adjustment phase, the training objectives consist of annotated data from the cold start stage and data sampled by the initialized reformulator.

2.3 Re-ranking Model

The task of the re-ranking model is to estimate a value p_{ik} that reflects the relevance of the retrieved document d_k to the query q_i. Similarly to previous work [33], we treat it as a binary classification task and use the BERT as the classifier, which feeds the "[CLS]" representation into an output layer to compute p_{ik}.

During the feedback adjustment phase, we obtain the fine-grained score s_{ij} by calculating the relevance between the original query q_i and the candidate documents. The score is based on a metric function, which is our ultimate goal. For example, for the *Precision@K* metric, we can calculate the score in the following way.

$$s_{ij} = 1/K * \sum_{k=1}^{K} p_{ik} \tag{4}$$

Note that other metrics can also be calculated in a similar way.

3 Experiments

3.1 Experimental Settings

Datasets. We conduct experiments on four datasets to validate the effectiveness of our framework, including MS MARCO, Robust04, ANTIQUE [15], and E-Commerce.

MS MARCO is constructed from Bing's search query logs and web documents retrieved by Bing. We follow previous studies [8,34,55] and use the standard data split.

Robust04 is constructed from TREC disks 4 and 5 and focuses on poorly performing topics. We follow a previous study [26] to use the five folds with three folds used for training, one fold for validation and the other fold for testing.

ANTIQUE is constructed from Yahoo! Webscope L6 and is mainly used for non-factoid question-answering research. We follow a previous work [15] to use the standard data split.

E-Commerce is constructed from one anonymous E-Commerce website. We annotate the query-document pairs using the query logs. We also collect a large number of unlabeled queries for further fine-tuning.

Detailed statistics of the datasets are shown in Table 1. Note that for Robust04 and ANTIQUE, we do not divide the queries into labeled and unlabeled sets, as the number of queries in these datasets is relatively low.

Table 1. Statistics of Datasets.

Dataset	#Doc	#Query	#Unlabeled Query	#Len
MS MARCO	8.8M	510k	322k	1.1k
Robust04	0.5M	249	–	0.7k
ANTIQUE	0.4M	2626	–	0.04k
E-Commerce	30M	282k	200k	0.1k

Evaluation Protocol. We follow previous studies [15,26,34] to employ the *Mean Reciprocal Rank (MRR)*, the *Precision (P)*, the *Normalized Discounted Cumulative Gain (NDCG)* and the *Recall (R)* metrics. For the E-Commerce dataset, we design a novel metric called *Re-ranking Precision (REKP)*. The *REKP* is the precision derived from the re-ranking model instead of the gold label, which provides a fine-grained perspective to evaluate the effectiveness of the results. We also adopt an online metric called *Good Ratio* to assess whether rewritten queries can be used for deployment, which is computed by a powerful discriminator.

Models. For the reformulator, we adopt T5$_{base}$ as the generator. For the re-ranking model, we use Roberta$_{base}$ [24] as the backbone of the classifier.

Baselines. We compare our methods with the following baselines that have been published over recent years.

Sparse Retrieval Methods. This group of methods provides sparse query and document representations for retrieval. Specifically, we report the experimental results of BM25 [41], BM25+RM3 [22], SDM (Sequential Dependence Model) [29], DeepCT (Deep Contextualized Term Weighting Model) [5], doc2query-T5 [34], COIL-tok (Contextualized Exact Match Retrieval Model) [10] and SPLADE (Sparse Lexical and Expansion Model) [8].

Dense Retrieval Methods. This group of methods learns dense query and document representations for retrieval. Specifically, we report the experimental results of Conv-KNRM (Convolutional Kernel-based Neural Ranking Model) [6], Vanilla BERT [33], CEDR-KNRM (Contextualized Embeddings for Document Ranking) [26], ANCE (Approximate nearest neighbor Negative Contrastive Estimation) [49], RocketQA [38], aNMM (Attention-Based Neural Matching Model) [51] and DRMM-TKS (Deep Relevance Matching Model) [12].

Other Methods. For the E-Commerce dataset, the T5QR [48] uses human rewrites to train a supervised QR model based on the pre-trained model T5. We propose a simple retrieval-based QR (RetQR) method that retrieves similar queries as the rewriting result using the search engine. Besides, we compare our method with the previous state-of-the-art retrieval-based query reformulation methods used in the e-commerce search engine, including Dense Retriever [16] that uses the dense retriever to retrieve similar queries, Swing [52] that is an effective graph-based QR method, Session2Vec that utilizes the context information in the session to construct query embeddings, and Item2Vec that leverages the user-item click information to build the query embeddings.

3.2 Main Results

Tables 2, 3, 4, and 5 show the main results. We observe that: 1) Our FRED method can be integrated with existing sparse retrieval methods and achieves significant performance gains over the original methods on all datasets. It demonstrates the effectiveness of our FRED method. 2) With a lower computational cost, our framework outperforms most of the dense retrieval methods and achieves performance comparable to those of previous state-of-the-art methods. Note that the proposed FRED framework focuses on the first-stage retrieval that aims to achieve high recall efficiently, while the Re-Ranker model focuses on the subsequent re-ranking stage. 3) Table 5 shows that compared to the traditional generation method T5QR, our FRED framework shows consistent performance improvements. On the contrary, the retrieval-based QR method performs poorly. This indicates that using the re-ranking model for filtering and feedback is crucial for the reformulator. 4) Our FRED framework can be trained in a semi-supervised manner, taking advantage of large amounts of unlabeled queries for further improvement in the first-stage retrieval. 5) For the E-Commerce dataset, the *Re-Ranker Pecision (REKP)* score and the *Precision (P)* score exhibit a positive correlation, which demonstrates that the re-ranking models can effectively evaluate the quality of rewritten queries. Moreover, the *REKP* significantly

Table 2. Performance on MS MARCO.

Model	MS MARCO	
	MRR@10	R@1000
ANCE [49]	0.33	0.959
RocketQA [38]	**0.37**	**0.979**
BM25 [41]	0.184	0.853
BM25+RM3 [22]	0.157	0.861
DeepCT [5]	0.243	0.913
doc2query-T5 [34]	0.277	0.947
COIL-tok [10]	**0.341**	0.949
SPLADE [8]	0.322	**0.955**
Re-Ranker (HLATR [55])	0.401	-
FRED–**BM25**	0.336	0.945
FRED–**BM25 (Semi-Sup)**	**0.343**	**0.959**
FRED–BM25+RM3	0.313	0.937
FRED–BM25+RM3 (Semi-Sup)	0.318	0.941

Table 3. Performance on Robust04.

Model	Robust04	
	P@20	NDCG@20
ConvKNRM [6]	0.3349	0.3806
DRMM-TKS [12]	0.3641	0.4135
CEDR-KNRM [26]	**0.4769**	**0.5475**
BM25 [41]	0.3123	0.4140
BM25+RM3 [22]	0.3462	0.4309
SDM [29]	**0.3749**	**0.4353**
Re-Ranker (Vanilla BERT)	0.4459	0.4917
FRED–**BM25**	0.3946	0.4463
FRED–**BM25+RM3**	**0.4012**	**0.4496**

surpasses the P across all methods, indicating that the majority of relevant documents in the dataset are unlabeled.

We also deploy 35 million rewritten queries on the E-Commerce Search Engine and compare our method with previous QR methods. The results are shown in Table 6, and we observe that our method significantly outperforms the state-of-the-art models on *Good Ratio* score by **+20.20%** with a comparable number of queries. This indicates that our framework can effectively alleviate the query drift problem (the underlying intent of the original query is changed during query reformulation).

Table 4. Performance on ANTIQUE.

Model	ANTIQUE			
	P@1	P@10	NDCG@1	NDCG@10
DRMM-TKS [12]	0.43337	0.3005	0.4949	0.4531
aNMM [51]	**0.4847**	**0.3306**	**0.5289**	**0.4904**
BM25 [41]	**0.3333**	**0.2485**	**0.4411**	**0.4334**
BM25+RM3 [22]	0.3192	0.2401	0.4287	0.4258
Re-Ranker (Vanilla BERT)	0.6998	0.4671	0.7015	0.6349
FRED–**BM25**	**0.5794**	**0.3683**	**0.5362**	**0.5071**
FRED–**BM25+RM3**	0.5513	0.3522	0.5196	0.5004

Table 5. Performance on E-Commerce.

Model	E-Commerce	
	REKP@100	P@100
Human Label	0.7072	0.0437
RetQR	0.4328	0.0214
T5QR [48]	**0.5470**	**0.0344**
FRED–**Search Engine**	0.8066	0.0706
FRED–**Search Engine (Semi-Sup)**	**0.8388**	**0.0780**

Table 6. Large-Scale Experimental results on E-Commerce.

Model	Good Ratio	#Avg. queries
Dense Retrieval [16]	0.2951	3.4141
Swing [52]	0.1281	1.5145
Session2Vec	0.0938	1.3637
Item2Vec	0.0571	1.932
FRED–**Search Engine (Semi-Sup)**	**0.4971**	1.9857

3.3 Ablation Study

We conduct an ablation study on Robust04 to validate the effectiveness of individual components of our proposed FRED framework. We compare our approach with the following methods: 1) **w/ T5-Large** denotes training the reformulator with $T5_{large}$ instead of $T5_{base}$. 2) **w/o Filter** means that we do not perform the filtering operation on the rewritten candidate queries. 3) **w/o Feedback** means that we remove the feedback adjustment stage in the training phase.

Table 7. Ablation results on Robust04.

Model	P@20	NDCG@20
FRED	0.3946	0.4463
w/ T5-Large	0.3960 (**+0.14%**)	0.4537 (**+0.74%**)
w/o Filter	0.3226 (**−7.20%**)	0.3874 (**−5.89%**)
w/o Feedback	0.3715 (**−2.31%**)	0.4257 (**−2.06%**)

Table 7 shows the ablation results on Robust04. We observe that: 1) Replacing the generator used in the reformulator with the larger pre-trained model does not yield significant improvements. This may be because the quality and quantity of the dataset are the bottleneck for the query rewriting task, rather than the scale of the model. 2) The filtering module can significantly improve performance. This is because the generation-based rewriting methods are unstable, and directly using the generated queries as the final results will greatly hurt the performance. 3) The feedback adjustment stage consistently improves performance. This is because the unlabeled and mislabeled problems in the dataset can be corrected by the powerful re-ranking model in this stage.

4   Related Work

4.1   Query Reformulation

There have been two lines of related work for query reformulation: (1) Global-based methods, typically refine a query based on global resources (e.g., Word-Net [11], thesauri [45], Wikipedia, and Dbpedia [1]). Those methods heavily rely on the quality of external resources. (2) Local-based methods, usually rewrite the query using the local relevance feedback. There are various methods to obtain the relevance feedback, including explicit feedback [42], implicit feedback [19], and pseudo-relevance feedback [22,25,42,44,47,54].

4.2   Pseudo-Relevance Feedback Methods

Traditional PRF methods expand queries by utilizing statistical information on the feedback documents [42]. Some studies [28,40] adopt probabilistic models and treat the top retrieved documents as relevant documents to improve the estimation of the model parameters. Some other works [18,22,37,53,54] try to build a query language model by utilizing the feedback set. However, these methods do not directly optimize retrieval performance [2,4,30,56], thus leading to poor retrieval performance. To address the above problems, LoL [57] designs a novel regularization loss to enable performance comparisons across various revisions. Some studies [32,35] propose a general reinforcement learning framework to optimize the retrieval performance. Unlike all the above methods, our method fully utilizes the powerful re-ranking model and the unlabeled data to improve the query reformulation to boost the retrieval performance.

4.3   Retrieval-Augmented Model

Recently, the retrieval augmented methods have been applied in various tasks such as Language Modeling [13,21], Dialogue Response Generation [31,46], and Neural Machine Translation [50]. These methods retrieve documents from a large corpus and then attend over these documents to help predict the final results, which is similar to the PRF method. Some methods [17,20] propose a pipeline method that trains the retriever and the reader independently. That is to say, the retriever network is trained in isolation from the reader network, which limits performance. To address the problem, some studies [14,23,43] propose a joint retriever and reader model that can be trained in an end-to-end fashion. Our method can also be viewed as a Retrieval Augmented method, since we use the retrieved documents as the context to generate the target queries.

5   Conclusion

In this paper, we propose a novel fine-grained feedback framework for query reformulation, which utilizes the powerful re-ranking model to provide more reasonable fine-grained feedback. We further propose a semi-supervised method to enhance the performance. The experimental results and analysis show that our framework is an effective solution for query reformulation with fewer data annotations and better generalization.

Acknowledgments. This work was supported by the NSFC (No. 61402403, No. 62072399), Zhejiang Provincial Natural Science Foundation of China (No. LZ23F020009), Chinese Knowledge Center for Engineering Sciences and Technology, MoE Engineering Research Center of Digital Library. Thanks to the Hangzhou AI Computing Center for providing the Atlas computing cluster.

References

1. Aggarwal, N., Buitelaar, P.: Query expansion using wikipedia and dbpedia. In: Forner, P., Karlgren, J., Womser-Hacker, C. (eds.) CLEF 2012 Evaluation Labs and Workshop, Online Working Notes, Rome, 17–20 September 2012. CEUR Workshop Proceedings, vol. 1178. CEUR-WS.org (2012)
2. Amati, G., Carpineto, C., Romano, G.: Query difficulty, robustness, and selective application of query expansion. In: McDonald, S., Tait, J. (eds.) ECIR 2004. LNCS, vol. 2997, pp. 127–137. Springer, Heidelberg (2004). https://doi.org/10.1007/978-3-540-24752-4_10
3. Brown, T.B., et al.: Language models are few-shot learners. arXiv preprint arXiv:2005.14165 (2020)
4. Croft, W.B., Harper, D.J.: Using probabilistic models of document retrieval without relevance information. J. Documentation **35**(4), 285–295 (1979). https://doi.org/10.1108/eb026683
5. Dai, Z., Callan, J.: Context-aware term weighting for first stage passage retrieval. In: Huang, J.X., et al. (eds.) Proceedings of the 43rd International ACM SIGIR Conference on Research and Development in Information Retrieval, SIGIR 2020, Virtual Event, 25–30 July 2020, pp. 1533–1536. ACM (2020). https://doi.org/10.1145/3397271.3401204
6. Dai, Z., Xiong, C., Callan, J., Liu, Z.: Convolutional neural networks for soft-matching n-grams in ad-hoc search. In: Chang, Y., Zhai, C., Liu, Y., Maarek, Y. (eds.) Proceedings of the Eleventh ACM International Conference on Web Search and Data Mining, WSDM 2018, Marina Del Rey, 5–9 February 2018, pp. 126–134. ACM (2018). https://doi.org/10.1145/3159652.3159659
7. Fani, H., Tamannaee, M., Zarrinkalam, F., Samouh, J., Paydar, S., Bagheri, E.: An extensible toolkit of query refinement methods and gold standard dataset generation. In: Hiemstra, D., Moens, M.-F., Mothe, J., Perego, R., Potthast, M., Sebastiani, F. (eds.) ECIR 2021. LNCS, vol. 12657, pp. 498–503. Springer, Cham (2021). https://doi.org/10.1007/978-3-030-72240-1_54

8. Formal, T., Piwowarski, B., Clinchant, S.: SPLADE: sparse lexical and expansion model for first stage ranking. In: Diaz, F., Shah, C., Suel, T., Castells, P., Jones, R., Sakai, T. (eds.) SIGIR '21: The 44th International ACM SIGIR Conference on Research and Development in Information Retrieval, Virtual Event, 11–15 July 2021, pp. 2288–2292. ACM (2021). https://doi.org/10.1145/3404835.3463098

9. Freitag, M., Al-Onaizan, Y.: Beam search strategies for neural machine translation. In: Luong, T., Birch, A., Neubig, G., Finch, A.M. (eds.) Proceedings of the First Workshop on Neural Machine Translation, NMT@ACL 2017, Vancouver, 4 August 2017, pp. 56–60. Association for Computational Linguistics (2017). https://doi.org/10.18653/v1/w17-3207

10. Gao, L., Dai, Z., Callan, J.: COIL: Revisit exact lexical match in information retrieval with contextualized inverted list. In: Toutanova, K., et al. (eds.) Proceedings of the 2021 Conference of the North American Chapter of the Association for Computational Linguistics: Human Language Technologies, NAACL-HLT 2021, Online, 6–11 June 2021, pp. 3030–3042. Association for Computational Linguistics (2021). https://doi.org/10.18653/v1/2021.naacl-main.241

11. Gong, Z., Cheang, C.W., Leong Hou, U.: Web query expansion by wordNet. In: Andersen, K.V., Debenham, J., Wagner, R. (eds.) DEXA 2005. LNCS, vol. 3588, pp. 166–175. Springer, Heidelberg (2005). https://doi.org/10.1007/11546924_17

12. Guo, J., Fan, Y., Ai, Q., Croft, W.B.: A deep relevance matching model for ad-hoc retrieval. In: Mukhopadhyay, S., et al. (eds.) Proceedings of the 25th ACM International Conference on Information and Knowledge Management, CIKM 2016, Indianapolis, 24–28 October 2016, pp. 55–64. ACM (2016). https://doi.org/10.1145/2983323.2983769

13. Guu, K., Hashimoto, T.B., Oren, Y., Liang, P.: Generating sentences by editing prototypes 6, 437–450 (2018). https://doi.org/10.1162/tacl_a_00030

14. Guu, K., Lee, K., Tung, Z., Pasupat, P., Chang, M.W.: REALM: retrieval-augmented language model pre-training. arXiv preprint arXiv:2002.08909 (2020)

15. Hashemi, H., Aliannejadi, M., Zamani, H., Croft, W.B.: ANTIQUE: a non-factoid question answering benchmark. In: Jose, J.M., et al. (eds.) ECIR 2020. LNCS, vol. 12036, pp. 166–173. Springer, Cham (2020). https://doi.org/10.1007/978-3-030-45442-5_21

16. Huang, P.S., He, X., Gao, J., Deng, L., Acero, A., Heck, L.P.: Learning deep structured semantic models for web search using clickthrough data. In: He, Q., Iyengar, A., Nejdl, W., Pei, J., Rastogi, R. (eds.) 22nd ACM International Conference on Information and Knowledge Management, CIKM'13, San Francisco, 27 October–1 November 2013, pp. 2333–2338. ACM (2013). https://doi.org/10.1145/2505515.2505665

17. Izacard, G., Grave, E.: Leveraging passage retrieval with generative models for open domain question answering. In: Merlo, P., Tiedemann, J., Tsarfaty, R. (eds.) Proceedings of the 16th Conference of the European Chapter of the Association for Computational Linguistics: Main Volume, EACL 2021, Online, 19–23 April 2021, pp. 874–880. Association for Computational Linguistics (2021). https://doi.org/10.18653/v1/2021.eacl-main.74

18. Jaleel, N.A., et al.: UMass at TREC 2004: novelty and HARD. In: Voorhees, E.M., Buckland, L.P. (eds.) Proceedings of the Thirteenth Text REtrieval Conference, TREC 2004, Gaithersburg, Maryland, 16–19 November 2004. NIST Special Publication, vol. 500–261. National Institute of Standards and Technology (NIST) (2004)

19. Joachims, T.: Optimizing search engines using click through data. In: Proceedings of the Eighth ACM SIGKDD International Conference on Knowledge Discovery and Data Mining, 23–26 July 2002, Edmonton, pp. 133–142. ACM (2002). https://doi.org/10.1145/775047.775067

20. Karpukhin, V., et al.: Dense passage retrieval for open-domain question answering. In: Webber, B., Cohn, T., He, Y., Liu, Y. (eds.) Proceedings of the 2020 Conference on Empirical Methods in Natural Language Processing, EMNLP 2020, Online, 16–20 November 2020, pp. 6769–6781. Association for Computational Linguistics (2020). https://doi.org/10.18653/v1/2020.emnlp-main.550

21. Khandelwal, U., Levy, O., Jurafsky, D., Zettlemoyer, L., Lewis, M.: Generalization through memorization: Nearest neighbor language models. In: 8th International Conference on Learning Representations, ICLR 2020, Addis Ababa, 26–30 April 2020. OpenReview.net (2020)

22. Lavrenko, V., Croft, W.B.: Relevance-based language models. In: Croft, W.B., Harper, D.J., Kraft, D.H., Zobel, J. (eds.) SIGIR 2001: Proceedings of the 24th Annual International ACM SIGIR Conference on Research and Development in Information Retrieval, 9–13 September 2001, New Orleans, pp. 120–127. ACM (2001). https://doi.org/10.1145/383952.383972

23. Lewis, P.S.H., et al.: Retrieval-augmented generation for knowledge-intensive NLP tasks. In: Larochelle, H., Ranzato, M., Hadsell, R., Balcan, M.F., Lin, H.T. (eds.) Advances in Neural Information Processing Systems 33: Annual Conference on Neural Information Processing Systems 2020, NeurIPS 2020, 6–12 December 2020, Virtual, pp. 9459–9474 (2020)

24. Liu, Y., et al.: RoBERTa: a robustly optimized BERT pretraining approach. arXiv preprint arXiv:1907.11692 (2019)

25. Lv, Y., Zhai, C.: A comparative study of methods for estimating query language models with pseudo feedback. In: Cheung, D.W.L., Song, I.Y., Chu, W.W., Hu, X., Lin, J. (eds.) Proceedings of the 18th ACM Conference on Information and Knowledge Management, CIKM 2009, Hong Kong, 2–6 November 2009, pp. 1895–1898. ACM (2009). https://doi.org/10.1145/1645953.1646259

26. MacAvaney, S., Yates, A., Cohan, A., Goharian, N.: CEDR: Contextualized embeddings for document ranking. In: Piwowarski, B., Chevalier, M., Gaussier, É., Maarek, Y., Nie, J.Y., Scholer, F. (eds.) Proceedings of the 42nd International ACM SIGIR Conference on Research and Development in Information Retrieval, SIGIR 2019, Paris, 21–25 July 2019, pp. 1101–1104. ACM (2019). https://doi.org/10.1145/3331184.3331317

27. Madaan, A., et al.: Self-refine: iterative refinement with self-feedback. arXiv preprint arXiv:2303.17651 (2023). https://doi.org/10.48550/arXiv.2303.17651

28. Maron, M.E., Kuhns, J.L.: On relevance, probabilistic indexing and information retrieval. J. ACM **7**(3), 216–244 (1960). https://doi.org/10.1145/321033.321035

29. Metzler, D., Croft, W.B.: A Markov random field model for term dependencies. In: Baeza-Yates, R.A., Ziviani, N., Marchionini, G., Moffat, A., Tait, J. (eds.) SIGIR 2005: Proceedings of the 28th Annual International ACM SIGIR Conference on Research and Development in Information Retrieval, Salvador, 15–19 August 2005, pp. 472–479. ACM (2005). https://doi.org/10.1145/1076034.1076115

30. Mitra, M., Singhal, A., Buckley, C.: Improving automatic query expansion. In: Croft, W.B., Moffat, A., van Rijsbergen, C.J., Wilkinson, R., Zobel, J. (eds.) SIGIR '98: Proceedings of the 21st Annual International ACM SIGIR Conference on Research and Development in Information Retrieval, 24–28 August 1998, Melbourne, pp. 206–214. ACM (1998). https://doi.org/10.1145/290941.290995

31. Moghe, N., Arora, S., Banerjee, S., Khapra, M.M.: Towards exploiting background knowledge for building conversation systems. In: Riloff, E., Chiang, D., Hockenmaier, J., Tsujii, J. (eds.) Proceedings of the 2018 Conference on Empirical Methods in Natural Language Processing, Brussels, 31 October–4 November 2018, pp. 2322–2332. Association for Computational Linguistics (2018). https://doi.org/10.18653/v1/d18-1255

32. Montazeralghaem, A., Zamani, H., Allan, J.: A reinforcement learning framework for relevance feedback. In: Huang, J.X., et al. (eds.) Proceedings of the 43rd International ACM SIGIR Conference on Research and Development in Information Retrieval, SIGIR 2020, Virtual Event, 25–30 July 2020, pp. 59–68. ACM (2020). https://doi.org/10.1145/3397271.3401099

33. Nogueira, R., Cho, K.: Passage re-ranking with BERT. arXiv preprint arXiv:1901.04085 (2019)

34. Nogueira, R., Lin, J., Epistemic, A.: From doc2query to docTTTTTquery. Online Preprint 6 (2019)

35. Nogueira, R.F., Cho, K.: Task-oriented query reformulation with reinforcement learning. In: Palmer, M., Hwa, R., Riedel, S. (eds.) Proceedings of the 2017 Conference on Empirical Methods in Natural Language Processing, EMNLP 2017, Copenhagen, 9–11 September 2017, pp. 574–583. Association for Computational Linguistics (2017). https://doi.org/10.18653/v1/d17-1061

36. Ouyang, L., et al.: Training language models to follow instructions with human feedback. In: Advances in Neural Information Processing Systems, pp. 27730–27744. Curran Associates, Inc. (2022)

37. Ponte, J.M., Croft, W.B.: A language modeling approach to information retrieval. In: Croft, W.B., Moffat, A., van Rijsbergen, C.J., Wilkinson, R., Zobel, J. (eds.) SIGIR '98: Proceedings of the 21st Annual International ACM SIGIR Conference on Research and Development in Information Retrieval, 24–28 August 1998, Melbourne, pp. 275–281. ACM (1998). https://doi.org/10.1145/290941.291008

38. Qu, Y., et al.: RocketQA: an optimized training approach to dense passage retrieval for open-domain question answering. In: Toutanova, K., et al. (eds.) Proceedings of the 2021 Conference of the North American Chapter of the Association for Computational Linguistics: Human Language Technologies, NAACL-HLT 2021, Online, 6–11 June 2021, pp. 5835–5847. Association for Computational Linguistics (2021). https://doi.org/10.18653/v1/2021.naacl-main.466

39. Raffel, C., et al.: Exploring the limits of transfer learning with a unified text-to-text transformer. J. Mach. Learn. Res. **21**, 140:1–140:67 (2020)

40. Robertson, S.E., Jones, K.S.: Relevance weighting of search terms. J. Am. Soc. Inf. Sci. **27**(3), 129–146 (1976)

41. Robertson, S.E., Zaragoza, H.: The probabilistic relevance framework: BM25 and beyond. Found. Trends Inf. Retr. **3**(4), 333–389 (2009). https://doi.org/10.1561/1500000019

42. Rocchio, J.J.: Relevance feedback in information retrieval. In: Salton, G. (ed.) The Smart Retrieval System - Experiments in Automatic Document Processing, pp. 313–323. Prentice-Hall, Englewood Cliffs (1971)

43. Sachan, D.S., Reddy, S., Hamilton, W.L., Dyer, C., Yogatama, D.: End-to-end training of multi-document reader and retriever for open-domain question answering. In: Ranzato, M., Beygelzimer, A., Dauphin, Y.N., Liang, P., Vaughan, J.W. (eds.) Advances in Neural Information Processing Systems 34: Annual Conference on Neural Information Processing Systems 2021, NeurIPS 2021, 6–14 December 2021, Virtual, pp. 25968–25981 (2021)

44. Salton, G., Buckley, C.: Improving retrieval performance by relevance feedback. J. Am. Soc. Inf. Sci. **41**(4), 288–297 (1990). https://doi.org/10.1002/(SICI)1097-4571(199006)41:4288::AID-ASI83.0.CO;2-H

45. Shiri, A.A., Revie, C.: Query expansion behavior within a thesaurus-enhanced search environment: a user-centered evaluation **57**(4), 462–478 (2006). https://doi.org/10.1002/asi.20319

46. Weston, J., Dinan, E., Miller, A.H.: Retrieve and refine: improved sequence generation models for dialogue. In: Chuklin, A., Dalton, J., Kiseleva, J., Borisov, A., Burtsev, M.S. (eds.) Proceedings of the 2nd International Workshop on Search-Oriented Conversational AI, SCAI@EMNLP 2018, Brussels, 31 October 2018, pp. 87–92. Association for Computational Linguistics (2018). https://doi.org/10.18653/v1/w18-5713

47. Willett, P.: Document Retrieval Systems. Taylor Graham Publishing (1988)

48. Wu, Z., et al.: CONQRR: conversational query rewriting for retrieval with reinforcement learning. In: Goldberg, Y., Kozareva, Z., Zhang, Y. (eds.) Proceedings of the 2022 Conference on Empirical Methods in Natural Language Processing, EMNLP 2022, Abu Dhabi, 7–11 December 2022, pp. 10000–10014. Association for Computational Linguistics (2022)

49. Xiong, L., et al.: Approximate nearest neighbor negative contrastive learning for dense text retrieval. In: 9th International Conference on Learning Representations, ICLR 2021, Virtual Event, 3–7 May 2021. OpenReview.net (2021)

50. Xu, J., Crego, J.M., Senellart, J.: Boosting neural machine translation with similar translations. In: Jurafsky, D., Chai, J., Schluter, N., Tetreault, J.R. (eds.) Proceedings of the 58th Annual Meeting of the Association for Computational Linguistics, ACL 2020, Online, 5–10 July 2020, pp. 1580–1590. Association for Computational Linguistics (2020). https://doi.org/10.18653/v1/2020.acl-main.144

51. Yang, L., Ai, Q., Guo, J., Croft, W.B.: aNMM: ranking short answer texts with attention-based neural matching model. In: Mukhopadhyay, S., et al. (eds.) Proceedings of the 25th ACM International Conference on Information and Knowledge Management, CIKM 2016, Indianapolis, 24–28 October 2016, pp. 287–296. ACM (2016). https://doi.org/10.1145/2983323.2983818

52. Yang, X., Zhu, Y., Zhang, Y., Wang, X., Yuan, Q.: Large scale product graph construction for recommendation in e-commerce. arXiv preprint arXiv:2010.05525 (2020)

53. Zhai, C., Lafferty, J.: Model-based feedback in the KL-divergence retrieval model. In: Tenth International Conference on Information and Knowledge Management (CIKM 2001), pp. 403–410 (2001)

54. Zhai, C., Lafferty, J.D.: Model-based feedback in the language modeling approach to information retrieval. In: Proceedings of the 2001 ACM CIKM International Conference on Information and Knowledge Management, Atlanta, 5–10 November 2001, pp. 403–410. ACM (2001). https://doi.org/10.1145/502585.502654

55. Zhang, Y., Long, D., Xu, G., Xie, P.: HLATR: enhance multi-stage text retrieval with hybrid list aware transformer reranking. arXiv preprint arXiv:2205.10569 (2022)

56. Zheng, Z., Hui, K., He, B., Han, X., Sun, L., Yates, A.: BERT-QE: contextualized query expansion for document re-ranking. In: Cohn, T., He, Y., Liu, Y. (eds.) Findings of the Association for Computational Linguistics: EMNLP 2020, Online Event, 16–20 November 2020. Findings of ACL, vol. EMNLP 2020, pp. 4718–4728. Association for Computational Linguistics (2020). https://doi.org/10.18653/v1/2020.findings-emnlp.424

57. Zhu, Y., Pang, L., Lan, Y., Shen, H., Cheng, X.: LoL: a comparative regularization loss over query reformulation losses for pseudo-relevance feedback. In: Amigó, E., Castells, P., Gonzalo, J., Carterette, B., Culpepper, J.S., Kazai, G. (eds.) SIGIR '22: The 45th International ACM SIGIR Conference on Research and Development in Information Retrieval, Madrid, 11–15 July 2022, pp. 825–836. ACM (2022). https://doi.org/10.1145/3477495.3532017

Retrieval-Augmented Document-Level Event Extraction with Cross-Attention Fusion

Yuting Xu[1], Chong Feng[1](✉), Bo Wang[1], Jing Huang[2,3,4], and Xinmu Qi[5]

[1] School of Computer Science and Technology, Beijing Institute of Technology, Beijing, China
{ytxu,fengchong,bwang}@bit.edu.cn
[2] Huajian Yutong Technology Co., Ltd., Beijing, China
[3] State Key Laboratory of Media Convergence Production Technology and Systems, Beijing, China
[4] Xinhua New Media Culture Communication Co., Ltd., Beijing, China
sklmcpts@xinhua.org
[5] Stony Brook Institute, Anhui University, Anhui, China

Abstract. Document-level event extraction intends to extract event records from an entire document. Current approaches adopt an entity-centric workflow, wherein the effectiveness of event extraction heavily relies on the input representation. Nonetheless, the input representations derived from earlier approaches exhibit incongruities when applied to the task of event extraction. To mitigate these discrepancies, we propose a Retrieval-Augmented Document-level Event Extraction (RADEE) method that leverages instances from the training dataset as supplementary event-informed knowledge. Specifically, the most similar training instance containing event records is retrieved and then concatenated with the input to enhance the input representation. To effectively integrate information from retrieved instances while minimizing noise interference, we introduce a fusion layer based on cross-attention mechanism. Experimental results obtained from a comprehensive evaluation of a large-scale document-level event extraction dataset reveal that our proposed method surpasses the performance of all baseline models. Furthermore, our approach exhibits improved performance even in low-resource settings, emphasizing its effectiveness and adaptability.

Keywords: Document-level event extraction · Retrieval-augmented · Cross-attention fusion

1 Introduction

Event extraction (EE) aims to detect event and extract event-related arguments from unstructured text. As a crucial subtask of information extraction, event extraction can provide valuable structured information for a range of downstream tasks, such as knowledge graph construction, question answering, etc. Previous studies mostly focus on sentence-level EE [2,8], while in real-world scenarios,

F. Wu et al. (Eds.): SMP 2023, CCIS 1945, pp. 218–229, 2024.
https://doi.org/10.1007/978-981-99-7596-9_16

event arguments usually scatter in multiple sentences. And there may be multiple events in a single document. Therefore, document-level EE has attracted increasing interest from both academic and industrial researchers. In contrast to sentence-level EE, document-level EE aims to extract events and their corresponding arguments from the entire document. This task is particularly challenging as it requires capturing long-range dependencies among scattered parameters in the document. Moreover, in a multi-event scenario, it's difficult to identify the intricate interactions between events and their arguments.

To handle these challenges, Zheng et al. [16] model the document-level EE task as an event table-filling task by generating an entity-based directed acyclic graph (DAG). Xu et al. [14] propose a heterogeneous graph to capture the global interactions between entities and sentences. To capture the interdependency among events, they memorize the extracted event records with a Tracker module. Zhu et al. [17] propose a non-autoregressive document-level EE model that extracts event records from pruned complete graphs based on entities in parallel. In an entity-centric workflow, the performance of event extraction is largely limited by the input representation. However, the input representation obtained by existing methods lacks of event-oriented knowledge and is not suitable enough for the event extraction task.

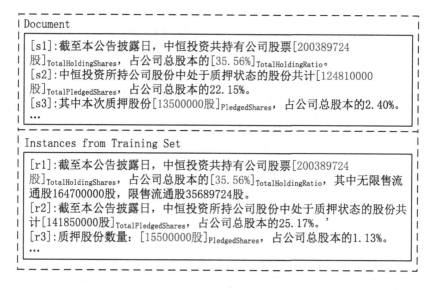

Fig. 1. An input example for document EE with retrieval instances

Based on our observation of financial documents, we find that texts containing the same type of event and arguments are very similar in terms of both semantics and syntax. As shown in Fig. 1, we have a document for input "...Among them, 13,500,000 shares were pledged, accounting for 2.40% of the total share capital of the Company..." and instances retrieved from training set

"...Number of pledged shares: 15,500,000 shares, accounting for 1.13% of the total share capital of the Company....". The retrieval instances contain the same event type *EquityPledge* and matching role type (*TotalHoldingShares, TotalHoldingRatio, TotalPledgedShares, PledgedShares*) as the original input document.

Motivated by the weakness of existing methods and our observation, we propose a retrieval-augmented document-level event extraction method that regards similar training instances which are event-aware as enhanced knowledge of the current input. First of all, we initialize the retrieval store with training set. Then we introduce a retriever module to get most similar instances for the current input. After that, We concatenate the retrieval instance with the current input as an augmented input and fed it into an unified encoder. Meanwhile, we propose a fusion layer to integrate the important information from retrieval instances and avoid disturbance of irrelevant features.

We conduct extensive experiments on ChFinAnn [16] dataset and the results demonstrate that our method outperforms all baselines, especially in precision. Besides, our method achieves strong performance in low resource setting.

2   Related Work

2.1   Document-Level Event Extraction

Compared with sentence-level EE, document-level EE requires extracting event records from the whole document. In general, most of the event information is likely to be concentrated in a central sentence of the document. Extracting arguments from the identified central sentence and filling in the remaining arguments from neighboring sentences can solve the document-level EE problem [15]. While zheng et al. [16] model document-level EE task as constructing an entity-based directed acyclic graph(DAG) with transformer for global information interaction. Xu et al. [14] propose a heterogeneous graph on the basic of [16] to capture the global interaction of sentences and entities and use a Tracker module to track predicted event records. In order to accelerate the training and inference of document-level EE, Zhu et al. [17] propose a parallel decoding method based on pruned complete graph instead of sequential decoding method based on DAG.

In contrast to above methods, our retrieval-augmented method utilizes event-informed training instances to enhance the input representation for document-level EE.

2.2   Retrieval-Augmented Methods

Retrieval-augmented methods have been demonstrated to be effective across various NLP tasks, such as open domain question answering [5], dialogue generation [1], etc. Most previous work utilizes retrieval text as templates to guide models in generating target text [10], or retrieves external knowledge to enhance language model pre-training [3]. However, Wang et al. [13] use the retrieval text as auxiliary information to enhance the semantic representation of the input sentence, and it's proven to be a simple and effective method. Since retrieving knowledge from external sources requires large scale indexing and is very time consuming, training data with the same data distribution as inference data is regarded as the retrieval source for supervised tasks [12]. To leverage the retrieval information, previous methods usually concatenate the input and retrieval text, and utilize an encoder to obtain a representation of the input tokens. Different from them, we employ a cross-attention mechanism to better integrate the retrieval knowledge into the representation of the input tokens.

3 Preliminaries

Before introducing our method, we first clarify several key concepts of document-level event extraction. 1) **entity mention**: a text span within a document that refers to an entity. 2) **event role**: a predefined field of event table. 3) **event argument**: an entity plays a specific role in an event record. 4) **event record**: a combination of arguments, each of which plays an event role, together form an entry of the event table.

Given a document composed of sentences $D = \{s_i\}_{i=1}^{|D|}$, where each sentence contains a sequence of words $s_i = \{w_{i,j}\}_{j=1}^{|s_i|}$, we let $\{r_i\}_{i=1}^{|D|}$ denote the retrieval training instances aligned to D. In our research on document-level event extraction tasks, our objective is to extract all event records from a given document. It's important to note that there may be more than one event record within a single document. Each event record encapsulates an event type and a set of entities that perform pre-defined roles.

4 Method

In this section, we describe our retrieval-augmented document-level event extraction method in Fig. 2. Firstly, sentences in the document are fed into the retriever module to get most similar instances, followed by an encoder and a fusion layer to obtain an augmented input representation (sec Sect. 4.1). Then a CRF [7] layer and an event type classifier are used to extract all the candidate entities in the sentences (sec Sect. 4.2) and identify the event types in the document (sec Sect. 4.3), respectively. Finally, the proposed method conducts event record generation non-autoregressively (sec Sect. 4.4).

4.1 Retrieval-Augmented Representation

Retriever. The retrieval store $\mathcal{RS}$ is initialized with data from the training set, and instances will be extracted from $\mathcal{RS}$ during training and inference for augmentation. As two sentences with high similarity may contain the same type of events and arguments, we split the document in the training set into sentences and initialize $\mathcal{RS}$ with these sentences and their corresponding mention labels which are annotated in the training set. All mentions of an entity share the same entity type label, the type set consists of all role types and four non-role types.

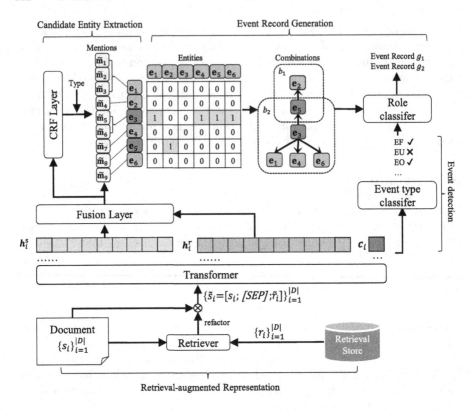

Fig. 2. Framework for Retrieval-Augmented Document-level Event Extraction

As the main component of the retriever, S-BERT [9] is used to retrieve instances from $\mathcal{RS}$ that are most similar to the input sentences. S-BERT uses Siamese BERT-Networks to obtain semantically meaningful sentence embedding, which can be used to calculate the similarity between sentences. Specifically, we first use S-BERT to encode sentences in $\mathcal{RS}$ into dense vectors as a retrieval key set $\mathbf{K}$. Meanwhile, for sentences in the input document, we regard the dense vectors obtained by S-BERT as the retrieval query set $\mathbf{Q}$. Then we rank the instances in $\mathcal{RS}$ according to the cosine-similarity scores of $\mathbf{Q}$ and $\mathbf{K}$. Finally, the most similar instance $r_i = (a_i, \{(m_{i,j}, l_{i,j})\}_{j=1}^{N_m^{(i)}})$ is retrieved for each input sentence s_i, where a_i denotes retrieval sentence, $m_{i,j}$ and $l_{i,j}$ denote mentions in a_i and their corresponding type labels.

To better leverage mention labels in the retrieval instance, a template of the form "$a_i \; m_{i,1} \; is \; l_{i,1} \;, ..., m_{i,N_m^{(i)}} \; is \; l_{i,N_m^{(i)}}$" is applied to refactor r_i into $\tilde{r}_i$. Here, a special token such as *[SEP]* is used to distinguish s_i and $\tilde{r}_i$. The original input sentence and the refactored instance are concatenated to construct the new retrieval-augmented input $\tilde{s}_i = [s_i; [SEP]; \tilde{r}_i]$.

Information Fusion Module. The Information fusion module consists of two parts: a unified encoder and a fusion layer. After getting the retrieval-augmented input $\tilde{s}_i$, we encode it with a Transformer [11] encoder layer:

$$\mathbf{h}_i = [\mathbf{h}_i^s, \mathbf{h}_i^r] = \text{Transformer}(\tilde{s}_i) \tag{1}$$

where $\mathbf{h}_i^s$ denotes the representation of the original input sentence, $\mathbf{h}_i^r$ denotes the representation of refactored retrieval instance.

It should be noted that only the token representation corresponding to the original input will be fed into the subsequent module. Instead of directly truncating the token representation of the input sentence, we use an additional fusion layer to integrate the information from the retrieval instance. In detail, a cross-attention mechanism is used to control the features to be focused on in the retrieval information. The cross-attention mechanism is used in the decoder of Transformer [11] to mix two different sequences of information. It is often used in machine translation or text image classification. After that, we employ a residual connection [4] to obtain the fused sentence representation:

$$\tilde{\mathbf{h}}_i = \text{softmax}(\frac{\mathbf{h}_i^s {\mathbf{h}_i^r}^T}{\sqrt{d_h}})\mathbf{h}_i^r + \mathbf{h}_i^s \tag{2}$$

where $\tilde{\mathbf{h}}_i$ is an augmented representation that incorporates important information from the retrieval instance.

4.2 Candidate Entity Extraction

We formulate the entity extraction task as a sentence-level sequence labeling task and use roles of matched arguments with classical BIO (Begin, Inside, Other) schema as entity labels. Given a retrieval augmented sentence representation $\tilde{\mathbf{h}}_i$, a Conditional Random Fields (CRF) [7] is adopted to identify entity mentions. For training, we minimize the following loss:

$$\mathcal{L}_{ner} = -\sum_{s_i \in D} \log p(y_i \mid s_i) \tag{3}$$

where s_i is the input sentence and y_i is the gold label sequence of s_i.

For all token-level representations in an entity mention, a mean-pooling operation is used to get mention representation $\mathbf{m}_k$. Besides, we obtain a mention type embedding $\mathbf{l}_k$ by looking up an type embedding table. We get the final mention representation $\tilde{\mathbf{m}}_k = \mathbf{m}_k + \mathbf{l}_k$ by concatenating the mention representation $\mathbf{m}_k$ and the mention type embedding $\mathbf{l}_k$. As an entity may have many mentions, we use a max-pooling operation to aggregate all the mentions of the j-th entity to obtain the entity representation $\mathbf{e}_j = \text{Max}(\{\tilde{\mathbf{m}}_k\}_{k \in e_j})$.

4.3 Event Detection

Event detection can be formulated as a multi-label binary classification task, since there may be multiple types of events in a single document. Based on the sentences representations $\mathbf{C} = [\mathbf{c}_i; ...; \mathbf{c}_{|D|}]$ obtained by encoder, we use an attention mechanism to obtain the document embedding $\mathbf{d}$ for each event type.

$$\mathbf{d} = \text{Attention}(\mathbf{Q}, \mathbf{C}, \mathbf{C}) \tag{4}$$

where $\mathbf{Q}$ is trainable parameter. Then, we stack a linear classifier over $\mathbf{d}$ to predict whether the event is identified. A binary cross entropy function is used to calculate $\mathcal{L}_{det}$.

4.4   Event Record Generation

After obtaining the entity representations and event types through the above module, we perform event record generation with the help of pseudo-trigger as previous method did [17]. This section mainly includes: event record partition and role classification.

Event Record Partition. Given a set of candidate entities $\{e_j\}_{j=1}^{N_e}$, we can group the entities in the same event record together. For each type of event, an importance metric [17] is used to select an argument with the highest importance score as the pseudo-trigger. The pseudo-trigger is regarded as the core of the event record, and all arguments related to it are in the same event record. Formally, given an event record $g = \{pseudo - trigger, arg1, arg2\}$, we set $< pseudo - trigger, arg1 >= 1, < pseudo - trigger, arg2 >= 1, < arg1, arg2 >= 0$, where 1 means two arguments are related and 0 means two arguments are not related. As shown in the Fig. 2, the entity relationship matrix $\mathbf{U}$ is obtained by calculating the dot scaled similarity between the candidate entities.

$$\mathbf{U}_{i,j} = \text{sigmoid}(\frac{\phi_1(\mathbf{e}_i)\phi_2(\mathbf{e}_j)^T}{\sqrt{d_h}}) \tag{5}$$

where ϕ_1 and ϕ_2 are linear projections. We construct the loss $\mathcal{L}_{part}$ of this module using a binary cross entropy function. When inferencing, the threshold $\lambda = 0.5$ is applied to determine whether $\mathbf{U}_{i,j}$ is 0 or 1. Based on $\mathbf{U}$, we can select the pseudo-triggers that have relationships with other entities and unite all their associated entities to form original combinations $\mathcal{B}$ of event arguments.

Role Classification. We construct a pre-defined role classifier for each type of event and get the role results for all arguments in each combination. In detail, given the predicted event type t, there is pre-defined role classifier corresponding to the type t, and for each combination of event arguments $b_i \in \mathcal{B}$, the role probability of each argument in b_i is calculated by a sigmoid function. The loss $\mathcal{L}_{role}$ is formulated by a binary cross entropy function.

4.5   Training and Inference

The final loss is a weighted sum of the losses from four subtasks.

$$\mathcal{L} = \alpha_1 \mathcal{L}_{ner} + \alpha_2 \mathcal{L}_{det} + \alpha_3 \mathcal{L}_{part} + \mathcal{L}_{role} \tag{6}$$

For both training and inference processes, we utilize the training set as the retrieval store. Specifically, during the training phase, if the retrieved instance aligns with the current input, we opt for the second most similar instance for augmentation purposes. In the decoding stage, when multiple event records share identical key roles, we retain only one of these records. Here, the key role is determined based on the dataset ontology.

5   Experiment

5.1   Dataset

We evaluate our method on ChFinAnn [16] dataset. ChFinAnn is a large-scale document-level event extraction dataset constructed from Chinese financial documents without triggers. It includes 32040 documents and 47824 event records, focusing on five event types: Equity Freeze (EF), Equity Repurchase (ER), Equity Underweight (EU), Equity Overweight (EO), and Equity Pledge (EP). We divide the dataset into train/dev/test sets in the ratio of 8/1/1.

5.2   Experiments Setting

For the input, we set the maximum sentence length and the maximum number of sentences in the document to 256 and 64, respectively. We use 12-layer BERT as the encoder and BERT-base-chinese[1] as the pre-trained language model. The hidden layer dimension and the mention type embedding dimension are 768 and 32, respectively. During training, we employ the Adam [6] optimizer with the learning rate 5e-4 and set the training batch size to 32. The weights in the loss function are 0.05,1.0 and 1.0. We train our model for 100 epochs and choose the best epoch with the highest F1 score on the development set to evaluate the test set.

5.3   Baselines and Metrics

Baselines. To evaluate the effectiveness of our method, we compare it to the following baselines. 1) **DCFEE** [15] contains two variants that extract only one event record and multiple event records from the document. 2) **Doc2EDAG** [16] proposes an entity-based path-expanding paradigm to solve the document-level event extraction task. 3) **GIT** [14] captures the interaction between entities and sentences with a heterogeneous graph and record the extracted event records with a tracker. 4) **PTPCG** [17] is an efficient model that benefits from a non-autoregressive decoding strategy.

Metrics. We use precision, recall, and F1 score to evaluate the performance of our method. Following [16]'s setting, for each predicted event record, we select a record with the same event type and the most matched arguments from the ground truth event records and calculate precision, recall, and F1 score by comparing arguments. All metrics are computed as micro-average values.

---

[1] BERT-base-chinese. https://huggingface.co/bert-base-chinese.

Table 1. Results between baselines and our method on the ChFinAnn dataset. The best results are **bolded** and the second results are underlined. †: LSTM encoder. ‡: Transformer encoder. ◇: results from [14]

Model	P (%)	R (%)	F1 (%)
DCFEE-O †◇	67.7	54.4	60.3
DCFEE-M †◇	58.1	55.2	56.6
Doc2EDAG ‡◇	80.3	75.0	77.5
GIT ‡◇	82.3	**78.4**	<u>80.3</u>
PTPCG †	82.0	76.2	79.0
PTPCG ‡	82.7	76.1	79.3
RADEE †	**86.1**	74.6	80.0
-*Fusion layer*	82.5	76.8	79.5
RADEE ‡	<u>84.7</u>	77.4	**80.9**
-*Fusion layer*	83.9	<u>77.6</u>	80.6

5.4   Main Results

Overall Performance. Table 1 shows the comparison of our method and baselines on the ChFinAnn [16] dataset. We observe that:

1) Overall, our method outperforms others in terms of precision and F1 score. Compared to PTPCG, our method achieves significant improvements in different encoder settings, 1.0 f1, and 1.6 f1 respectively. Our transformer-based model outperforms PTPCG in all three metrics, which suggests the great effectiveness of our proposed method. Compared to other baselines, our method improves 2.0 precision and 0.6 micro F1 score. This indicates that using similar event-informed instances can enhance the input representation and thus improve document-level event extraction. However, we have a lower recall than GIT because its tracker module exploits interdependency information between event records.

2) The fusion layer incorporates important information from retrieval instances into the input representation. To explore its effect, we remove the fusion layer from our method. As shown in Table 1, our method experiences a decrease in precision by 0.8. This indicates that the fusion layer enables the model to focus on the important information from the retrieval instances while disregarding noise.

Cross-Sentence Records Scenario. Approximately 98% of the event records in the dataset are cross-sentence records. To evaluate the effectiveness of our method in the cross-sentence scenario, we divide the test set into four equal parts based on the number of sentences involved in the event records (from I to IV with increasing number of sentences involved). As shown in Table 2, our method shows the best results in almost every set, which suggests that our method can well overcome the cross-sentence issue, with the help of retrieval event-informed knowledge (Table 3).

Table 2. F1 scores on four sets with growing average number of involved sentences for records (increases from I to IV). ◇: results from [14]

Model	I	II	III	IV	Avg
DCFEE-O †◇	64.6	70.0	57.7	52.3	61.2
DCFEE-M †◇	54.8	54.1	51.5	47.1	51.9
Doc2EDAG ‡◇	79.6	82.4	78.4	72.0	78.1
GIT ‡◇	81.9	85.7	80.0	**75.7**	80.8
PTPCG ‡	82.6	84.5	79.4	73.0	79.9
RADEE ‡	**84.3**	**85.8**	**80.7**	75.4	**81.6**

Table 3. F1 scores on Single (S.) and Multi (M.) sets. EF/ER/EU/EO/EP refer to specific event types. ◇: results from [14]

Model	EF		ER		EU		EO		EP		Overall	
	S	M	S	M	S	M	S	M	S	M	S	M
DCFEE-O †◇	55.7	38.1	83.0	55.5	52.3	41.4	49.2	43.6	62.4	52.2	69.0	50.3
DCFEE-M †◇	45.3	40.5	76.1	50.6	48.3	43.1	45.7	43.3	58.1	51.2	63.2	49.4
Doc2EDAG ‡◇	79.7	63.3	90.4	70.7	74.7	63.3	76.1	<u>70.2</u>	84.3	69.3	81.0	67.4
GIT ‡◇	81.9	**65.9**	<u>93.0</u>	71.7	**82.0**	<u>64.1</u>	<u>80.9</u>	**70.6**	<u>85.0</u>	**73.5**	<u>87.6</u>	**72.3**
PTPCG ‡	<u>82.7</u>	61.9	<u>93.0</u>	**74.4**	79.9	63.4	**81.7**	63.3	<u>85.0</u>	71.1	<u>87.6</u>	69.8
RADEE ‡	**83.3**	<u>63.9</u>	**95.6**	<u>72.8</u>	<u>80.1</u>	**65.7**	80.4	67.9	**87.3**	<u>72.3</u>	**89.6**	<u>71.2</u>

Single and Multiple Scenarios. We divide the test set into two parts: *Single* containing only one event record and *Multi* containing multiple event records. Our method performs better on *Single* and is weaker than GIT overall on *Multi*. We attribute that GIT benefits from the interdependency between event records captured by an additional tracker module. Nevertheless, Table 1 illustrates that our method outperforms GIT overall, and is 2.4 times faster than GIT. Compared

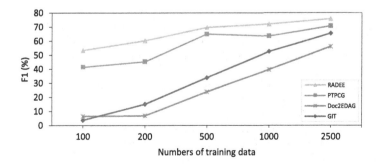

Fig. 3. F1 scores on ChFinAnn by different numbers of training data (100, 200, 500, 1000, 2500).

to PTPCG, our method has an improvement on both *Single* and *Multi* for almost every event type, with the final F1 score improving by 2.0 and 1.4, respectively.

Low Resource. Due to the high annotation cost, document-level EE suffers from high-quality annotated data scarcity. To verify the effectiveness of our method in low resource scenario, we sample 100, 200, 500, 1000, and 2500 documents from the training set and evaluate them on the origin test set. All baselines use the Transformer encoder. As shown in Fig. 3, our method outperforms all the baselines, with the largest improvement in the case of extreme lack of data. We believe that similar training instances provide the model with event information and event patterns, which can alleviate the problems caused by data scarcity.

6   Conclusion

In this work, we propose a retrieval-augmented document-level event extraction method with cross-attention fusion. The aim is to utilize event-informed training instances as auxiliary knowledge, facilitating the enhancement of input representation in the proposed approach by capturing inter-entity interactions effectively. We introduce a cross-attention fusion layer designed to aggregate pertinent information from retrieved data while mitigating the influence of extraneous noise. Experimental results demonstrate that our method effectively improves performance on the document-level event extraction dataset. In addition, the model demonstrates notable performances in low-resource settings, underscoring the efficacy of incorporating supplementary retrieval event-informed instances.

Acknowledgment. This work was supported by No. MKS20220076 and Institute of Science and Development, Chinese Academy of Sciences (No. GHJ-ZLZX-2023-04). We would like to thank the anonymous reviewers for their valuable comments and suggestions.

References

1. Cai, D., Wang, Y., Bi, W., Tu, Z., Liu, X., Shi, S.: Retrieval-guided dialogue response generation via a matching-to-generation framework. In: Proceedings of the 2019 Conference on Empirical Methods in Natural Language Processing and the 9th International Joint Conference on Natural Language Processing (EMNLP-IJCNLP) (2019)
2. Du, X., Cardie, C.: Event extraction by answering (almost) natural questions. In: Proceedings of the 2020 Conference on Empirical Methods in Natural Language Processing (EMNLP) (2020)
3. Guu, K., Lee, K., Tung, Z., Pasupat, P., Chang, M.W.: REALM: retrieval-augmented language model pre-training. In: Proceedings of the 37th International Conference on Machine Learning (2020)
4. He, K., Zhang, X., Ren, S., Sun, J.: Deep residual learning for image recognition. In: 2016 IEEE Conference on Computer Vision and Pattern Recognition (CVPR) (2016)

5. Izacard, G., Grave, E.: Leveraging passage retrieval with generative models for open domain question answering. In: Proceedings of the 16th Conference of the European Chapter of the Association for Computational Linguistics: Main Volume (2021)
6. Kingma, D.P., Ba, J.: Adam: a method for stochastic optimization. In: 3rd International Conference on Learning Representations, ICLR 2015, San Diego, CA, USA, 7–9 May 2015, Conference Track Proceedings (2015)
7. Lafferty, J.D., McCallum, A., Pereira, F.C.N.: Conditional random fields: probabilistic models for segmenting and labeling sequence data. In: Proceedings of the Eighteenth International Conference on Machine Learning (ICML 2001), Williams College, Williamstown, MA, USA, 28 June–1 July 2001 (2001)
8. Nguyen, T.H., Cho, K., Grishman, R.: Joint event extraction via recurrent neural networks. In: Proceedings of the 2016 Conference of the North American Chapter of the Association for Computational Linguistics: Human Language Technologies (2016)
9. Reimers, N., Gurevych, I.: Sentence-BERT: sentence embeddings using Siamese BERT-networks. In: Proceedings of the 2019 Conference on Empirical Methods in Natural Language Processing and the 9th International Joint Conference on Natural Language Processing (EMNLP-IJCNLP) (2019)
10. Su, Y., Wang, Y., Cai, D., Baker, S., Korhonen, A., Collier, N.: Prototype-to-style: dialogue generation with style-aware editing on retrieval memory. IEEE/ACM Trans. Audio Speech Lang. Process. **29**, 2152–2161 (2021)
11. Vaswani, A., et al.: Attention is all you need. In: Advances in Neural Information Processing Systems (2017)
12. Wang, S., et al.: Training data is more valuable than you think: a simple and effective method by retrieving from training data. In: Proceedings of the 60th Annual Meeting of the Association for Computational Linguistics (Volume 1: Long Papers) (2022)
13. Wang, X., et al.: Improving named entity recognition by external context retrieving and cooperative learning. In: Proceedings of the 59th Annual Meeting of the Association for Computational Linguistics and the 11th International Joint Conference on Natural Language Processing (Volume 1: Long Papers) (2021)
14. Xu, R., Liu, T., Li, L., Chang, B.: Document-level event extraction via heterogeneous graph-based interaction model with a tracker. In: Proceedings of the 59th Annual Meeting of the Association for Computational Linguistics and the 11th International Joint Conference on Natural Language Processing (Volume 1: Long Papers) (2021)
15. Yang, H., Chen, Y., Liu, K., Xiao, Y., Zhao, J.: DCFEE: a document-level Chinese financial event extraction system based on automatically labeled training data. In: Proceedings of ACL 2018, System Demonstrations (2018)
16. Zheng, S., Cao, W., Xu, W., Bian, J.: Doc2EDAG: an end-to-end document-level framework for Chinese financial event extraction. In: Proceedings of the 2019 Conference on Empirical Methods in Natural Language Processing and the 9th International Joint Conference on Natural Language Processing (EMNLP-IJCNLP) (2019)
17. Zhu, T., et al.: Efficient document-level event extraction via pseudo-trigger-aware pruned complete graph. In: Proceedings of the Thirty-First International Joint Conference on Artificial Intelligence, IJCAI-22 (2022)

Author Index

Printed in the United States
by Baker & Taylor Publisher Services